Governing Environmental Conflicts in China

Environmental conflicts are the source of many large-scale popular protests in China, with some protests substantially endangering social order. Such protests have often prompted severe counter measures by both national and local government, but have often then gone on to result in compromises whereby the demands of protesters have been largely met. This book considers the nature of environmental conflicts in China and the way in which the authorities have handled the situations. It includes detailed case studies of particular conflicts, relates the governance of environmental conflicts in China to wider discussions on the nature of governance and examines under what conditions government in China makes compromises. The book concludes by assessing the lessons for the future.

Yanwei Li is an Associate Professor in the School of Public Administration, Nanjing Normal University, China.

China Policy Series
Series Editor
Zheng Yongnian
East Asian Institute, National University of Singapore

Governing Environmental Conflicts in China

Yanwei Li

Routledge
Taylor & Francis Group

LONDON AND NEW YORK

First published 2018
by Routledge
2 Park Square, Milton Park, Abingdon, Oxon OX14 4RN

and by Routledge
605 Third Avenue, New York, NY 10017

First issued in paperback 2021

Routledge is an imprint of the Taylor & Francis Group, an informa business

Publisher's Note
The publisher has gone to great lengths to ensure the quality of this reprint but points out that some imperfections in the original copies may be apparent.

British Library Cataloguing-in-Publication Data
A catalogue record for this book is available from the British Library

Library of Congress Cataloging-in-Publication Data
A catalog record for this book has been requested

Typeset in Times New Roman
by Wearset Ltd, Boldon, Tyne and Wear

ISBN 13: 978-1-138-55190-9 (hbk)
ISBN 13: 978-0-367-43850-0 (pbk)

Book DOI: 10.4324/9781315147895

Contents

Tables

Preface

After over 30 years of fast economic development, many citizens in urban areas of China have accumulated substantial wealth and are starting to pursue other social values, such as justice, quality of life, public participation, transparency, and environmental quality. In this context, environmental conflicts increasingly occur when citizens oppose the planning, construction, and/or operation of a large number of industrial facilities. The top-down approach to coping with citizens' concerns that the Chinese state normally adopts does not work well in dealing with environmental conflicts involving diverse actors with various value priorities. What makes such conflicts even more intractable is the uncertainty about, and hence distrust of, new techniques. No scientific knowledge is available to convince citizens that these techniques are harmless to public health.

In Chapter 1, the main research question is presented. The main research question in this book is: *What is the nature and what are the causes of strategies applied by local governments during environmental conflicts concerning the planning, construction, and operation of industrial plants in urban China from 2007 to 2013?* This book is potentially interesting for scholars of governance and environmental issues around the world as it provides insights into how environmental conflicts are governed in a Chinese context and into how the "China paradox" can be explained. This book is of methodological interest as well, as it provides ideas about how to employ three case study strategies in one book: single case study, comparative case study, and qualitative comparative analysis (QCA). Last but not least, this book is of relevance to decision makers in China. It offers answers about how environmental conflicts should be governed and the best practices for this.

In Chapter 2, a literature review is made. In general, the topic studied in this book relates to four strands of literature: (1) on environmental conflicts; (2) on governance, decision making, and policy change; (3) on public participation in planning; and (4) on social conflict resolution in China. These four strands all provide insights on three issues: the nature of environmental conflicts, the strategies applied by Chinese local governments in environmental conflicts, and the explanation for the application of government strategies. What the four strands reveal about these issues are reviewed and reported in detail in Chapter 2.

In Chapter 3, a conceptual framework is constructed to describe and explain government strategies in governing Chinese environmental conflicts. It includes three key concepts: the policy game, government strategies, and the conditions that explain the application of government strategies. The *policy game* concept is used to analyze processes of environmental conflict resolution. The second concept, *government strategy*, is used to describe and categorize the concrete government actions that emerge in environmental conflicts. Six different government strategies, namely, go-alone, suppression, tension reduction, giving in, collaboration, and facilitation, are identified to function as an analytic tool to identify and categorize government actions in environmental conflicts. The third key concept is made up of *seven conditions that explain the application of government strategies*. They are: scale of protest, form of protest, position of higher-level governments, position of the national mass media, stage of projects, involvement of activists, and occurrence of events. The second main element of a conceptual framework is the relationship between the key concepts. Two approaches are used in this book to establish the relationships of the three key concepts: propositions and configurational thinking. Propositions are used to show how theoretically the individual conditions influence the application of government strategies. Configurational thinking allows us to explore the causal relationships between combinations of the conditions and the application of government strategies.

In Chapter 4, the research method and the research strategy are elaborated in detail. The research question established in this book relates to how environmental conflicts are governed in China and why particular strategies are applied by local governments to govern such conflicts. A qualitative case study is a better option to answer this type of question than quantitative approaches, as it can provide in-depth knowledge about how a specific outcome or process occurs. Combined application of three different case study strategies – single case study, comparative case study, and QCA – makes it possible to draw robust conclusions about the explanation of government strategies in environmental conflicts. Single case studies can provide in-depth knowledge about a case. They reveal how shifts in government strategies during environmental conflicts can be explained and how individual conditions shape the application of, or shifts in, government strategies. A limitation of single case studies is that their conclusions cannot be generalized. The comparative case study is a good option to remedy this. However, comparative case studies cannot be used to study how combinations of conditions influence the application of government strategies during environmental conflicts. Consequently, QCA is a good option.

In Chapter 5, a first single case study is reported: the Panyu waste incineration power plant. This case occurred in Guangzhou, an economically well-developed region of China. Five government strategies identified in Chapter 3, namely go-alone, suppression, tension reduction, and facilitation/collaboration were adopted by local governments during this case. I used process tracing to explain why local governments in Guangzhou applied these strategies over time. Finally, some conclusions are drawn, resulting in the confirmation, disconfirmation, or specification of some propositions drawn in Chapter 3.

In Chapter 6, a second single case study is reported: the Dalian PX case. This case occurred in Dalian, a traditional industrial base in China. Many key industrial plants have been constructed and operate there. Dalian Municipality had a strong economic reliance on local industrial plants because of the plants' enormous contributions to tax revenue. Consequently, Dalian Municipality had a strong incentive to align with local industries rather than take the interests of local citizens seriously. In this case, five government strategies were applied by Dalian Municipality: go-alone, suppression, tension reduction, giving in, and again tension reduction. Again, process tracing was used to explain these.

In Chapter 7, a first comparative case study is reported. The interest is *which conditions are important in explaining government strategies.* To this end, three patterns of government strategies in governing environmental conflicts are first identified by distinguishing three substantive outcomes regarding the debated industrial plants: project continuation, project cancellation, and project relocation. In this chapter, both the method of agreement and the method of difference are used to compare the three patterns of government strategies. First, the method of agreement is used to identify which conditions are relatively important in explaining the same pattern of government strategies. Second, the method of difference is used to identify which conditions are crucial in explaining the differences in patterns of government strategies.

In Chapter 8, a second comparative study uses csQCA (crisp-set QCA). One of the most promising advantages of QCA is that it allows researchers to make systematic and structured comparison, from which robust conclusions can be drawn about how the combinations of conditions lead to an outcome. In this book, csQCA is used to study how the combinations of four conditions, namely, scale of protest, form of protest, position of the central government, and stage of the project, can lead to various government strategies in environmental conflicts. Specifically, two outcomes are studied: the occurrence and the nonoccurrence of government compromises.

In Chapter 9, some main findings in this book are summarized. First, answers to the three sub-questions raised in Chapter 1 are presented. The second main aim is to reflect upon the contributions of this book in three dimensions: theoretical, practical, and methodological. Regarding the limitations of this book, it is suggested that the conceptual framework should be further elaborated, more empirical data are required, and the dynamic dimensions of the cases should be studied. The third issue resolved in Chapter 9 relates to the research agenda.

Acknowledgments

Comments and suggestions on various iterations of this book were generously provided by Joop Koppenjan, Martin de Jong, Vincent Homburg, Victor Bekkers, Meine Pieter van Dijk, Ernst ten Heuvelhof, Jurian Edelenbos, Harry Geerlings, Benjamin van Rooij, Peter Ho, Ank Michels, Ellen van Bueren, Stijn Verbeek, Stephane Moyson, Ziya Aliyev, Eva Thomann, Barbara Vis, Maarten Vink, Yi Liu, Yijia Jing, Stefan Verweij, Arthur Edwards, Markus Haverland, Ruth Prins, Lasse Gerrits, Ruth Post, Jolien Grandia, Iris Korthagen, Anne Annink, Diana Giebels, Rianne Dekker, Joris van der Voet, Lieselot Vandenbussche, Stephan Dorsman, Wouter Spekkink, Ingmar van Meerkerk, Danny Schipper, Jasper Eshuis, Wenting Jiao, Wei Sun, Qiaomei Yang, Qiong Gong, Danyang Du, Ruxi Wang, Qiushi Liang, Wenqi Dang, Hongchun Zhang, Yihong Liu, Xiaoli Lv, and Da Chi.

My thanks go too to those who gave me help during my fieldwork in China. I want to express my thanks to Professor Kaifeng Yang, Professor Ying Liang, Dr. Lei Zhang, Professor Zongchao Peng, Professor Fanbin Kong, Professor Jianguo Zhou, Professor Kechang Wu, Professor Fang Wang, Professor Weiqing Guo, and Dr. Guoliang Shi, for their help in my fieldwork in Nanjing, Beijing, Shanghai, and Guangzhou. I also want to express my thanks to all the interviewees.

This book would not have been able to be published if there was no support from Professor Yongnian Zheng and Mr. Peter Sowden. Thanks for their generous help and constructive comments. I also want to express my gratitude to Catherine O'Dea for editing the whole manuscript. Her work really made my book more readable and attractive.

Finally, thanks to my parents and my wife. My parents always encouraged me to be a dream-seeker. Their encouragement was invaluable for me in finishing this book. My wife, Qinqin Jiang, has never complained about my choice as a researcher. Her support made it possible for me to work on this book with few disruptions.

April 5, 2017
In Nanjing

1 Government strategies in environmental conflicts

A governance issue for Chinese local governments

Citizens in China nowadays are not that surprised by the occurrence of environmental conflicts, because they have happened in several cities, such as Xiamen (2007), Beijing (2007), Dalian (2011), Ningbo (2012), and Kunming (2013). One notable characteristic shared by all these incidents is that citizens who assemble together to protest against the construction of industrial facilities are arguably motivated by the presumed negative impacts of these facilities on environmental and public health. Occasionally, these mobilized protests came as a surprise to local governments, which immediately had to take action to deal with them, such as temporarily halting the project, project relocation, repression using state force, or face-to-face talks. Citizens, however, sometimes did not give up easily. They continued their opposition through letters and visits, seeking help from mass media or experts, or even initiating a new round of protests. The governments had to continuously adjust their approach to cope with these various strategies applied by citizens.

The episodes described above are a general representation of the main phenomenon studied in this book, namely, how conflicts concerning the planning, construction, and operation of industrial facilities are governed by the Chinese state.

Problem statement: environmental conflict as a governance issue in China

Nowadays, sustainable development is a leading aspiration of humankind. A collection of new ideas and concepts, such as the sustainable city, eco-city, smart-city, innovative city, green lifestyle, and sustainable built environment are booming and increasingly institutionalized in the world (de Jong, Wang, and Yu 2013; Newman and Jennings 2008; Register 2006). China, as the biggest developing country, seems to reflect a complex scenario regarding how to handle the relationship between sustainability and economic development. Since 1978, it has made great economic advances. The order of social values around China was clear: the economy first and all other values second. Against that background, the Chinese state developed its economy at the expense of the environment. However, the situation has changed. On the one hand, the Chinese state has to continue to prioritize economic development because China still needs some time to become a developed country. On the other hand, Chinese citizens increasingly have diverse policy preferences (Mertha 2009). Other social values, such as environmental quality, social justice, and quality of life, cannot be ignored as much as they were in the past.

Among these preferences, environmental quality has become an important concern for Chinese citizens. It has been reported that 90 percent of underground water in cities and 70 percent of rivers are polluted; indeed, a third of these rivers are so toxic that they endanger health. Seven of the world's 10 most polluted cities are in China, and smog contributes to around a million premature deaths each year. China is the world's largest emitter of carbon dioxide, and choking smog in cities reaches levels that make it hazardous to go outside.[1] According to the Ministry of Environmental Protection of China's figure for 2010,[2] the cost of pollution has reached about 1.5 trillion RMB (Renminbi), or roughly 3.5 percent of GDP. For the Chinese state, one unanticipated outcome caused by degrading environmental quality is the occurrence of a large number of environmental conflicts. Since 1997, environmental conflicts have increased annually by 25 percent (Hou and Zhang 2009). In recent years, there is increasing evidence that the upcoming middle class in urban regions has started to express their concerns and disagreements with government decisions, which adopt a rule-based approach; national laws, policies, or regulations are cited to remedy perceived governmental mismanagement (Johnson 2010). One consequence of this is that the Chinese state often has to reinforce public participation to accommodate this. In addition, in contrast to many violent protests that have occurred in rural China in response to environmental degradation, these citizens work within the existing political system as much as possible. They attempt to depoliticize their protest actions by establishing transparency and public consultation as their *master frame* (Johnson 2016), which is a generic type of collective action frame (Snow and Benford 1992). In contrast to other types of master frame, such as human rights, political reform, or political freedom, this master frame is relatively acceptable for the Chinese state, which creates an environment for negotiation between it and its citizens (Johnson 2016).

In general, these new features of protests initiated by Chinese citizens in relation to environmental issues cause a dilemma for the state. China, like many other authoritarian countries, has established concession and repression as two crucially important strategies to cope with social conflicts (Cai 2010). However, both suppression and concession appear to be ineffective in coping with environmental conflicts in contemporary China. First, the occurrences of environmental conflicts with mobilized protests potentially endanger social order, the top priority of the Chinese state. Actions must be taken to end them as soon as possible. Second, the depoliticized nature of the protests initiated by citizens makes the use of suppression inappropriate as it will significantly damage state legitimacy. Third, making concessions to meet citizens' demands by enhancing public participation does not match the authoritarian nature of the state. It can be thus concluded that the governance of environmental conflicts is a challenging issue for the Chinese state; concession or suppression cannot help it out of environmental conflicts. The state needs to adopt a greater variety of strategies to govern environmental conflicts in order to achieve more satisfactory outcomes. Therefore, it is highly important to study how the Chinese state governs environmental conflicts. This book aims to explore this issue. Specifically, it studies how environmental conflicts in the Chinese context can be understood, how they are governed in practice by the Chinese state, and how the evolvement of government strategies can be explained.

Problem analysis and knowledge gaps

To start with, a preliminary literature review was undertaken in order to identify knowledge gaps and possible research questions. A literature review is presented in Chapter 2. The following gaps were identified:

- *Knowledge gap 1: The nature of environmental conflicts in China*: Studies on the nature of environmental conflicts are not rare (see Bingham 1986; Dukes 2004; Glavovic, Dukes, and Lynott 1997; Huys and Koppenjan 2009). These conflicts can be defined as confrontations among disputants who are fighting over natural resources, such as land, gas, water, or oil. They may also be viewed as wicked problems without a unified formulation of problems and solutions (Rittel and Webber 1973). These problems are not only of a technical nature: the application of new technologies to address environmental problems may result in unanticipated dangers and risks, and various actors debate with one another (Fischer 2000). Furthermore, environmental conflicts may be seen as governance issues. Citizens, for example, may be dissatisfied because they are excluded from decision-making processes, and they demand transparency, openness, and involvement. To sum up, we know a lot about environmental conflicts, but not about environmental conflicts in China.
- *Knowledge gap 2: The strategies applied by the Chinese state to cope with environmental conflicts*: Some scholars (see Amy 1987; Glasbergen 1995; Susskind and Cruikshank 1987; van Bueren, Klijn, and Koppenjan 2003) have contributed to the issue of how environmental conflicts can be governed. Their general conclusion is that collaboration, negotiation, or facilitation are the best practices for the governance of environmental conflicts. Various actors involved should negotiate with one another in order to jointly make decisions. The state can function as an independent third party to facilitate interactions among various actors in order to advance the conflict resolution process. This approach draws primarily on governance practices in Western democracies. Little, however, is known about the governance of environmental conflicts in the Chinese context. In addition, some scholars (see Cai 2004, 2008a, 2008b, 2010; Deng and Yang 2013; O'Brien and Deng 2015; O'Brien and Li 2006; Shi and Cai 2006) have researched how

non-state actors in China apply various strategies to shape government decisions during social conflicts. How Chinese governments cope with environmental conflicts has not been well explored. This is the second knowledge gap.

* *Knowledge gap 3: The explanation of government strategies during Chinese environmental conflicts*: Some public policy and governance scholars (see Baumgartner and Jones 1993; Cobb and Elder 1983; Kingdon 2010) have extensively discussed government decisions in Western democracies. These studies provide insights into the explanation of government policies. Having studied social conflicts in China, some scholars (see Cai 2002, 2010; O'Brien and Li 2006; Shi and Cai 2006) specifically study under what conditions non-state actors succeed in achieving their goals. But little research has been done on explaining the emergence or application of strategies by Chinese governments.

Environmental conflicts may arise with regard to various types of industrial facilities, such as chemical factories, dams, battery plants, nuclear power plants, and the high-speed railways (Li, Liu, and Li 2012; He, Mol, and Lu 2016). In this book, environmental conflicts relate mainly to the planning, construction, and operation of two types of industrial plants: waste incineration power plants and PX plants. Nowadays in China, these two types of industrial projects are most widely debated. To study them is a good starting point to explore how the Chinese state – especially local governments at the municipal level – governs environmental conflicts. This book will contribute to bridging the above three knowledge gaps by answering the following main research question:

* *What is the nature and what are the causes of strategies applied by local governments during environmental conflicts concerning the planning, construction, and operation of industrial plants in urban China from 2007 to 2013?*

The following sub-questions will be addressed in order to answer the main research question.

The first element of the main research question is about the nature of environmental conflicts in China; this corresponds to the first knowledge gap. To study the nature of Chinese environmental conflicts, the first sub-question is developed as follows:

* *What are the general characteristics of environmental conflicts in China? (What are environmental conflicts? Who are involved in environmental conflicts? What are their strategies? How do environmental conflicts evolve? What are the outcomes of environmental conflicts?)*

The second element of the main research question is about government strategies in environmental conflicts; this corresponds to the second knowledge gap. To study government strategies in environmental conflicts, the second sub-question is formulated as follows:

- *Which kinds of government strategies are applied by Chinese local governments in environmental conflicts?*

The third element of the main research question is about the explanation of the application of government strategies in environmental conflicts; this corresponds to the third knowledge gap. To research it, the third sub-question is elaborated as follows:

- *How can the application of strategies by local governments during environmental conflicts be explained?*

The above three sub-questions are the key issues to be addressed in this book.

Research relevance

The research questions having been established, the next step is to explain the relevance of studying the governance of environmental conflicts in China. This is structured in three aspects: theoretical, methodological, and practical relevance. They are elaborated as follows.

Theoretical value: advancing the development of governance theories

Since the 1960s, many citizens have opposed industrial plants in Western democracies. Planners and governments sometimes had to adapt their previous decisions in order to accommodate citizens' demands. Afterwards, scholars from various scientific backgrounds, such as planning, governance, public policy, social movement, conflict resolution, and the management of natural resources, started to study how environmental conflicts have evolved and been resolved (Amy 1987; Bingham 1986; Fischer 2000; Fischer and Forester 1993; Fisher, Ury, and Patton 2011; Glasbergen 1995; Koppenjan and Klijn 2004; Susskind and Cruikshank 1987). This book studies how environmental conflicts are governed in the Chinese context and this may contribute new theoretical insights to the governance field. It is thus highly relevant for governance and environmental scholars, especially those who specialize in the governance of environmental conflicts around the world. An intriguing issue for them is: can experiences in China with regard to the governance of environmental conflict be used elsewhere?

This book is also relevant for governance scholars who are curious about what is called the "China paradox": high economic development as well as improvement in social welfare but a relatively low score on government quality in all the commonly used measures (Fukuyama 2013; Rothstein 2015). There is no authoritative explanation for this. Some governance scholars argue that the responsive nature of Chinese governance may have one important explanation: the Chinese state often adjusts its policies in order to accommodate public opinion (Grabosky 2013; Hammond 2013). This is coined *responsive authoritarianism* (van Rooij, Stern, and Fürst 2014; Weller 2012). The Chinese state delivers stability, order, economic development, and the improvement of citizens' well-being in return for its continued control of social governance (Tsang 2009). Following this argument, the China paradox seems to make sense. However, responsive authoritarianism may not be the only explanation of the China paradox. Rothstein (2015) has found that the cadre type of public administration in the Chinese context may be more efficient than the Weberian model. This book will provide insights into the relationships between responsive authoritarianism and the China paradox. Does responsive authoritarianism exist in Chinese governance? If it exists, a relevant issue is: does it contribute to the explanation of the China paradox?

Methodological relevance: a multi-method approach to case study research

This book is a piece of relevant and interesting work for scholars who are specifically interested in qualitative case studies. In recent years, some methodologists have made efforts to improve the validity and robustness of qualitative case studies (see Baumgartner 2013; Beach and Pedersen 2013; Blatter and Haverland 2012; Brady and Collier 2010; Dul et al. 2010; George and Bennett 2005; Gerring 2007; Ragin 1987). Different from a lot of case study research that mostly only uses either a single case approach or a comparative study of a limited number of cases, this book combines three case study strategies: single case study, comparative case study, and qualitative comparative analysis (QCA). Single case studies are able to provide in-depth insights into complex phenomena of interest. However, their generalization is contested. Comparative case studies are helpful to structure comparison and identify the crucially important conditions that explain the outcomes that we are studying. One of their limitations is that they do not tell us how the combinations of conditions influence the outcomes in which we are interested. QCA is a very appropriate approach to make a systematic and structured comparison of a medium-N of cases and make inferences based on the causality of necessity and sufficiency (Ragin 1987). Moreover, it is useful to explore how, compared to single and comparative case studies, the combinations of conditions lead to specific outcomes. Its use also has drawbacks, however, such as the limited number of cases and of conditions to be studied. From a methodological perspective, the combined use of the three methods can offer more encompassing and robust answers to the research questions. This book will provide insights into how these three methods are applied in a combined way. It offers guidelines for researchers who are planning to do research projects using the qualitative case study method.

Practical relevance: providing policy guidelines for the state and insights for practitioners

This book is relevant for policymakers in China because it provides suggestions on how environmental conflicts should be governed. Environmental conflicts pose a great challenge for the Chinese state, which has a strong intention to govern them in a satisfactory way. Since the 1960s, many governments in Western democracies have experienced the same struggles, and they have suffered from these even in recent years (McAdam and Boudet 2012). This implies that they may have accumulated a lot of practical experience regarding the governance of environmental conflicts. This book will discuss them as they are useful in offering practical guidelines for decision makers in China, specifically local governments, regarding the question of how environmental conflicts can be governed.

Structure of this book

This book is organized in nine chapters. In Chapter 2, four general strands of literature associated with environmental conflicts and their resolution are systematically reviewed: (1) the environmental conflict literature; (2) the governance, the decision-making, and the policy change literature; (3) the public participation in planning literature; and (4) the social conflict resolution literature in China. The conceptual framework formulated on the basis of this literature review is elaborated in Chapter 3. The research methodology and research strategy are introduced in Chapter 4. Two single case studies, the Panyu case and the Dalian case, are respectively reported in Chapter 5 and Chapter 6. In Chapter 7, an in-depth comparative case study is reported with the aim of explaining the similarities and differences of the patterns of government strategies, using the method of agreement and the method of difference. In Chapter 8, a crisp-set qualitative comparative analysis (csQCA) is used to compare 10 cases of environmental conflicts in order to explore under what conditions Chinese local governments make compromises with local communities. In Chapter 9, conclusions, reflections, recommendations, and a future research agenda for conducting empirical research about the governance of environmental conflicts are presented and discussed.

Notes

1 www.aljazeera.com/indepth/opinion/2014/07/china-pollution-protests-2014729105 632310682.html, available on August 25, 2015.
2 www.cfr.org/china/chinas-environmental-crisis/p12608, available on August 25, 2015.

2 Government strategies in environmental conflicts

A literature review

Introduction

This chapter reviews relevant literature providing insights into government strategies in environmental conflicts. It is structured as follows. The first section introduces the literature-search strategy used, and four strands of literature are identified. In the following sections, these four strands are reviewed in order to address three issues: the nature of environmental conflicts, government strategies in environmental conflicts, and the conditions underlying these strategies. In the final section, some general conclusions are drawn.

Literature-search strategy and findings

In order to locate related literature about the topic *government strategies in environmental conflicts in China*, two search strategies were employed. First, several rounds of expert consultation with experts from Dutch universities led to the identification of four relevant strands of literature: (1) literature on environmental conflicts, (2) literature on governance, decision making, and policy change, (3) literature on public participation in planning, and (4) literature on social conflict resolution in China. The first three strands relate to the international community and the fourth specifically relates to China. It is plausible first to review the most relevant literature about environmental conflicts and their resolution. After all, an environmental conflict can be seen as a problem involving disagreements among different actors, and its resolution is a governance issue. In addition, literature on policy change may provide insights into the dynamics of government strategies and their explanation. Literature on governance, decision making, and policy change makes up the second strand of theory from which we use insights and concepts. The degree of participation may be one way to characterize government strategies in environmental conflicts, as various degrees of public participation imply different types of government strategies. The public participation in planning literature is therefore established as one source for review in this book. It is the third most relevant strand of literature on environmental conflicts and their resolution. The literature on the resolution of social conflicts in China can provide insights into the governance of environmental conflicts in China. Second, I used Google Scholar and the mega search engine of the Erasmus University Rotterdam, sEURch, which includes various databases (including ABI/Inform Complete, Econlit, JSTOR, Science-Direct, Social Science Citation Index, Web of Knowledge, and Web of Science) to identify related literature. Keyword-based computerized searches were employed to discover any such literature. Keywords included *conflict, environmental conflict resolution, social conflict resolution, environmental governance, governance strategy, government response, public participation in planning, environmental conflicts, policy change, conflict resolution in China*, and *environmental conflict in China*.

In Chapter 1, three research sub-questions were posed. The first sub-question is mainly about the nature of environmental conflicts. The second relates to the strategies applied by the Chinese state in governing environmental conflicts. The third focuses on the explanation of the application of (government) strategies in environmental conflicts. Thus, four strands of literature are reviewed to provide insights into the three issues of interest, specifically relating to the governance of environmental conflicts in China: the nature of environmental conflicts, (government) strategies to govern environmental conflicts, and the explanation of why particular government strategies are applied.

In the next four sections, the four strands of literature are reviewed sequentially: the literature on the nature of environmental conflicts, then the strategies applied to govern environmental conflicts, and finally the explanation of why particular government strategies are applied.

Environmental conflict literature

In 1962, a plan proposed by the Consolidated Edison Company of New York to build a pumped storage hydroelectric project at Storm King Mountain on the Hudson River in Cornwall initiated a long-lasting conflict around its environmental impacts. After 16 years of court battles, the involved parties finally resolved their differences in 1981, but the poor functioning of the traditional administrative and legal mechanisms to resolve environmental conflicts came in for substantial criticism. Since then, environmental conflict resolution (ECR), as an alternative to traditional environmental conflict resolution mechanisms, has been increasingly developed and practiced in the US. It is regarded as a voluntary process of environmental conflict resolution, in which the disputants negotiate with each other in order to achieve a solution that is acceptable for both sides.

The nature of environmental conflicts

According to Felstiner, Abel, and Sarat's (1980) study, the emergence and transformation of a conflict will follow the route from naming and framing to claiming. At first, citizens identify the seriousness of a harmful incident, with the transformation of an unperceived injurious experience to a perceived injurious experience. A perceived injurious experience will then be transformed into a grievance, and some individuals or social entities are blamed for this. Lastly, citizens with grievances will search for remedial actions from those responsible for their injuries.

In the 1990s, two research groups, the Environment and Conflicts Project (ENCOP) group headed by Günther Baechler and Kurt Spillmann, and the Toronto group headed by Thomas Homer-Dixon, provided their definitions of environmental conflicts. The former views environmental conflicts as various social, economic, ethical, religious, or territorial conflicts induced by environmental degradation, whereas the latter regards environmental conflicts as conflicts induced by environmental scarcity (Homer-Dixon 1999). In general, both definitions view environmental conflicts as conflicts induced by environmental degradation or scarcity.

Glavovic, Dukes, and Lynott (1997, 271) conclude that environmental disputes

> can thus be distinguished by virtue of their primary concern with the allocation and use of land, air, water, and living resources. This focus is manifested in disputes characterized by high levels of uncertainty and complexity with consequences that affect the public good, involve multiple stakeholders, and are subject to incongruous boundary conditions.

In the same fashion, Dukes (2004) defines an environmental conflict as a kind of public conflict relating to issues such as health, race, and governance. It may include a combination of these issues.

In addition, the environmental conflict literature reveals a lot about the characteristics of environmental conflicts. Susskind and Ozawa (1984) propose that environmental conflicts are highly complex, often heavily relying on technical data and analysis, and involve unpredictable interests (such as the interests of future generations) and considerable *externalities*. Susskind and Weinstein (1980) acknowledge some properties of environmental conflicts compared to conflicts in labor relations. These include (1) irreversibility: implying that damage to the environment or ecology cannot be reversed; (2) high indetermination: meaning that the nature, boundaries, stakeholders, and costs cannot be clearly established; (3) labeling as being done in the name of the public interest: signifying that more than one party may claim to represent the public interest; and (4) difficulty of implementation: implying that agreements are difficult to implement.

Three other features have been identified by scholars as characterizing environmental conflicts: their subjective, political, and risky nature.

1   *The subjective nature of environmental conflicts*: This refers to the fact that values, beliefs, and principles are involved in environmental conflicts. They cannot be empirically tested, and the involved stakeholders have underlying moral concerns about how the environment should be treated (Campbell and Floyd 1996).

2   *The political nature of environmental conflicts*: Environmental conflicts may be policy stalemates related to general policy debates about the allocation of natural resources (Amy 1987; Bingham 1986; O'Leary and Bingham 2003). In addition, they may occur during decision making or implementation processes at different levels of government or by diverse government agencies (O'Leary, Nabatchi, and Bingham 2005).

3   *The risky nature of environmental conflicts*: Environmental conflicts involve substantial risks. Although planners or governments make policy-oriented risk assessments about technical risks, citizens may still psychologically anticipate that harm or catastrophes may follow (Fischer 1990). For example, many citizens may worry about the health risks posed by the emissions released by the process of waste incineration (Rootes and Leonard 2009).

To summarize, the environmental conflict literature provides many insights into the definition and characteristics of environmental conflicts.

Government strategies for environmental conflict resolution

Before the early 1970s, American citizens had few options to address environmental conflicts. They could express their concerns through administrative hearings, judicial arbitration, litigation, or referendums (Dukes 2004; Susskind and Ozawa 1984; Susskind and Weinstein 1980). Sometimes, they might even take confrontational actions involving contentious public hearings, angry media volleys, or large-scale demonstrations (Susskind and McKearnan 1999). These traditional approaches, however, mostly failed to reconcile the conflicting claims of the involved parties, tending to generate less-than-optimal outcomes (Amy 1987).

In the late 1960s and the 1970s, environmental conflict in the US became a national concern (Glavovic, Dukes, and Lynott 1997). The search to achieve a satisfactory solution led to the emergence of a new approach for resolving environmental conflicts, termed environmental conflict resolution (ECR). Its emergence signified a revolutionary change in the practices of environmental conflict resolution, and it is viewed as less costly, less contentious (or adversarial), and more creative compared to the traditional judicial approach to environmental conflict resolution (Dorius 1993). This new approach is highly consensus-building oriented and negotiation based (O'Leary and Raines 2001; Orr, Emerson, and Keyes 2008). Negotiation is simply bargaining – a process of discussion and give-and-take among disputants who want to find a solution to a common problem (O'Leary, Nabatchi, and Bingham 2005, 190). Mostly, the negotiation process is open and flexible, and the affected stakeholders with diverging perceptions have direct and face-to-face discussions (Dukes 2004). Often, it seeks consensus rather than the use of majority rule as the basis for agreement, and it aims to enhance the affected participants' mutual understanding of the nature and the resolution of problems that they face

(Daniels and Walker 2001). Compared to formal legal processes, the negotiation approach makes it easier to arrive at tailor-made settlements suited to the various specific situations, enabling the transition from win–lose confrontations to joint problem-solving efforts. The outcomes of negotiation are not the lowest common denominator of the affected stakeholders' interests. Rather, an attempt is made to integrate various interests by seeking a compromise agreement that partially satisfies the objectives of all the affected stakeholders (Amy 1983). Moreover, the outcomes are perceived by the affected disputants to be efficient, fair, wise, and stable (Susskind and Cruikshank 1987). Some authors publish books specifically offering empirical analysis to illustrate the values of negotiation in environmental conflict resolution, including *Resolving Environmental Disputes: A Decade of Experience* (Bingham 1986), *Environmental Dispute Resolution* (Bacow and Wheeler 1984), *Mediating Land Use Disputes Pros and Cons* (Susskind, van der Wansem, and Ciccarelli 2000), and *The Politics of Environmental Mediation* (Amy 1987).

In general, two forms of negotiation for environmental conflict resolution are identified by scholars: non-assisted negotiation and assisted negotiation (Forester 1987; Susskind and Cruikshank 1987). Non-assisted negotiation implies that only the principal affected stakeholders are involved in voluntary and inclusive consensus-seeking processes. It is emergent and spontaneous. Stakeholders engage in face-to-face negotiation with one another in the hope of addressing their differences and searching for joint net gains (Susskind and Weinstein 1980). They mostly attempt to achieve their own interests, but they are willing to reconsider the perceptions and positions of others to explore ways of capitalizing on overlapping interests (Glavovic, Dukes, and Lynott 1997). If the involved stakeholders have been trained in negotiation, they may possess the skills to transform the adversarial relationships among disputants into side-by-side conflict resolution.

During environmental conflict resolution, when the affected stakeholders encounter stalemates in exploring possible win–win solutions, the involvement of third parties is necessary. This is called *assisted, arranged, managed,* or *mediated* negotiation (Susskind and Ozawa 1984), or *mediation* (Amy 1983). Assisted negotiation is a voluntary process in which the involved stakeholders jointly resolve their differences with the help of a third party. It is a deliberate design in essence. Unlike arbitrators, mediators or facilitators do not have the authority to impose a settlement. Rather, their main responsibility lies in their ability to assist the parties in resolving their differences (Susskind and Weinstein 1980). Of course, it is not only public sector entities that can mediate; entities in the independent private sector can also do so. In the US, a large number of skilled mediators from private companies and non-profit organizations are actively involved in environmental conflict resolution (Susskind and Cruikshank 1987). They function as neutral or impartial third parties to design rules or institutions to resolve differences among disputants in the hope of seeking consensus and achieving a mutually beneficial outcome (Nabatchi 2007). Many specific strategies can be applied by mediators to resolve environmental conflicts, such as reframing, setting the location of meetings, engaging in joint learning toward forming common ground, imposing deadlines, and proposing agendas for discussions (Forester 2006).

Some scholars emphasize that neither assisted nor non-assisted negotiation are a panacea for environmental conflicts. Often, they complement rather than replace the traditional approaches to environmental conflict resolution (Susskind and McKearnan 1999). Not all environmental conflicts can be resolved by negotiation, which is mostly used when there are controversies and differences in values among disputants (Susskind and Weinstein 1980). Furthermore, the consensus-oriented negotiation approach to environmental conflict resolution is criticized. One crucial attack is that environmental negotiation is a form of political control that exploits the disadvantaged groups who are thinking that they are being treated fairly (Amy 1987). If agreement based on consensus is the primary goal of environmental negotiation, the facilitators may shy away from the hard work necessary to understand disputants' concerns so as not to endanger the potential for agreement (Gregory, McDaniels, and Fields 2001). As a result, justice may be sacrificed to reach a jointly acceptable solution through the abuse of discretion in consensus-seeking processes (O'Leary and Husar 2002). What is more, consensus does not mean that all disputants will favor one alternative, and not all elements of the agreement may need support from all of them. Rather, they should see their values reflected in the same alternative (Gregory 2000).

In conclusion, from the literature on environmental conflict, three general strategies in terms of environmental conflict resolution can be derived, namely, the traditional strategy, the non-assisted negotiation strategy, and the assisted negotiation strategy. These three general strategies are options for all disputants, including governments, during environmental conflicts. It is concluded therefore that these three strategies can be adopted by governments during environmental conflicts. Their definitions are presented in Table 2.1.

Table 2.1 Government strategies regarding environmental conflict resolution

Strategy	*Definition*
Traditional	This strategy often emphasizes the importance of formal rules and procedures in resolving environmental conflicts. It mostly results in a winner-takes-all outcome.
Non-assisted negotiation	This strategy implies that governments negotiate with the other stakeholders during environmental conflicts. Mostly, it is an informal negotiation among the involved actors in the hope of achieving consensus-based solutions.
Assisted/mediated negotiation	This strategy often signifies that an acceptable independent party works as a mediator to design processes or institutions to facilitate the resolution of environmental conflicts.

Explanation of why particular strategies are applied in environmental conflicts

In general, few studies have been conducted to specifically explain the application of actors' strategies during environmental conflicts. In one study, Susskind and Weinstein (1980) argue that two factors push the disputants in environmental conflicts to give up the traditional litigation or judicial approach. One is the cost of contentiousness. Environmental conflicts imply high costs for the involved stakeholders, business persons, environmental NGOs (nongovernmental organizations), or the public, such as carrying costs (total cost of holding inventory) for large-scale land development projects, opportunity costs for delayed facilities, and construction costs of facilities. Consequently, the perceived high costs of stalemates resulting from confrontations may require the disputants to resolve their environmental conflicts out of court. Non-assisted or assisted negotiations are two of their options. The second factor is dissatisfaction with the traditional approach to environmental conflict resolution, which often tends to result in Pyrrhic victories (or lose-lose situations), and this accelerates the need for a consensual approach by the disputants.

Governance, decision-making, and policy change literature

The literature on governance, decision making, and policy change provides many insights into the nature of problem solving, government strategies for problem solving, and the explanation of why particular government strategies are applied to solve problems. The concepts *problem* and *conflict* are highly interrelated: some problems are characterized by high consensus about their nature and resolution, whereas others inherently involve disagreements among actors. Disagreements may evoke conflicts. In governance and decision-making processes, disagreements regarding the distribution of resources or positions may result in conflicts (Cobb and Elder 1983). In traditional policy analysis, problem solving is seen as an intellectual design. Governments use the command-and-control approach to impose solutions informed by policy analysis. However, problem solving is not a linear intellectual process. Rather, it is characterized as highly unpredictable in a multi-actor context in which various actors interact with one another and jointly influence it. The insights into problem solving derived from the governance, decision-making, and policy change literature are useful for studying the nature of conflict, government strategies for conflict resolution, and the explanation of the application of government strategies for conflict resolution.

Understanding the nature of conflicts

Problems are the ultimate focus of governance and policy sciences. Scholars have discussed the nature of problems (Hisschemöller and Hoppe 1995; Hoppe 2011a, 2011b; Radford 1977; Rittel and Webber 1973). Some of them view problems as troubling conditions characterized as unacceptable gaps between normative ideals or aspirational levels and present conditions (Dery 1984). Other scholars argue that problems do not exist "out there." Rather, they are analytical constructs or conceptual entities (Weiss 1989; Wildavsky 1979).

On the basis of two dimensions, agreement or disagreement on scientific knowledge and normative values, Hoppe (2002) differentiates four types of problems: technical problems, untamed technical problems, political problems, and wicked problems.

1   *Technical problems*: These are characterized as having a high degree of consensus both on scientific knowledge and social values. Road maintenance is a case in point.
2   *Untamed technical problems*: These are problems with high disagreements about the application of certain technologies and a high degree of agreement on social values. For example, everyone thinks that the Ebola virus is a problem that should be addressed, but no technology can be used to achieve this.
3   *Political problems*: These are referred to as problems that are technically certain but with little consensus on social values. The abortion issue is an example of such a problem. The application of this technique is socially controversial.
4   *Wicked problems*: These are characterized by high levels of disagreement on both social values and scientific knowledge. For example, the use of waste incineration techniques is hotly debated because their effects on environmental and public health are uncertain. In addition, waste incineration is not socially accepted as the dominant way to dispose of waste.

Two generic models of problem resolution can be identified. One is the mono-actor model and the other is the multi-actor model (Kickert, Klijn, and Koppenjan 1997). In the mono-actor model, governments and their decisions are at the center. Problem solving is viewed as a rational intellectual design, distinguishable in various linear phases: problem definition, problem analysis, identification of solutions, comparison of costs and benefits, selection, implementation, and evaluation (Anderson 1984; Hogwood and Gunn 1984; Parsons 1995). This implies that the nature of the problem is first established and then scientific knowledge is employed to design measures for its solution. Finally, an implementation program is developed, and the outcome is evaluated on the basis of the objectives established in the policy-making stage.

Some theories, models, or concepts, such as the multiple-stream theory (Kingdon 2010), the multi-actor decision-making theory (Allison 1971; Sabatier 1986), and the interactive decision-making theory (Edelenbos 2005; Klijn and Koppenjan 2000a; Torfing et al. 2012), share a common multi-actor perspective on problem solving. This multi-actor perspective assumes that various actors with diverse perceptions, interests, and strategies are involved in problem solving. In addition, no authoritative value hierarchy exists in the multi-actor context. These features make problems complex, and it is difficult for the involved actors to build a consensus about their nature and resolution.

Often, scholars regard this multi-actor model as the governance network model (Klijn and Koppenjan 2016; Rhodes 1996; Scharpf 1997; Sørensen and Torfing 2007). Since the pioneering work of O'Toole (1997) who called for networks to be treated seriously, the *network* concept has nowadays become fashionable. A network is mostly referred to as "multiorganizational arrangements for solving problems that cannot be achieved, or achieved easily, by single organizations" (Agranoff and McGuire 2001, 296). Moreover, it is widely used by scholars in different academic fields (for reviews, see Ansell and Gash 2008; Berry et al. 2004; Börzel 1998; Dowding 1995; Isett et al. 2011; Klijn and Koppenjan 2012; Lecy, Mergel, and Schmitz 2014; Lewis 2011; Provan, Fish, and Sydow 2007; Robinson 2006; van Waarden 1992). Generally, two different perspectives on networks are identified: the first views a network as a multi-actor institutional context in which problem solving occurs, and the second regards a network as a normative approach to problem solving. These two perspectives are elaborated as follows:

1   *Network as a multi-actor institutional context characterized by interactions of multiple actors*: The first network perspective assumes that problem solving is complex, chaotic, and messy, mostly like a non-linear and zigzag game involving various actors with diverging perceptions, objectives, interests, and strategies (Scharpf 1997; Teisman 2000). A network is an institutional context for complex problem solving, public service delivery, or policy implementation (Crozier and Friedberg 1980; Gage and Mandell 1990; Jordan 1990; Klijn and Skelcher 2007; Rhodes 1990, 2007). Resources in the network context are distributed among various actors, and no actor can easily achieve its own goal alone. The actors involved in the network context are thus assumed to be dependent on one another. From this network perspective, the problem-solving processes may experience several rounds following the interactions of various actors. In addition, some network characteristics, such as rules (formal or informal), patterns of interactions, divisions of resources, or trust, all influence the problem-solving processes (Klijn and Koppenjan 2000b).

2   *Network as a normative governance (steering or governing) approach*: In the work of some scholars (Kooiman 1993; Lewis 2011), network implies a specific form of governance mode. Network governance or collaborative governance means that governments are no longer supreme. It can be defined as a governing mode where one or more public agencies directly engage non-state stakeholders in a collective decision-making and implementation process. In addition, this governance model is formal, consensus-oriented, and deliberative (Ansell and Gash 2008). The network approach can also be regarded as a social steering approach, implying that governments tend to use a horizontal, rather than a hierarchical, approach for decision making and implementation (van Kersbergen and van Waarden 2004).

The two concepts, *problem* and *conflict*, are highly related to each other. What the governance, decision-making, and policy change literature tells us about problems and problem solving can shed light on the nature of conflicts and conflict resolution. The above two perspectives about networks have theoretical and normative implications for the nature of conflicts. The first perspective sees networks as the institutional context in which conflicts occur. The second perspective views network as a normative governing approach to conflict resolution, which corresponds with the perspective that regards network as a governance strategy (different from the government strategy and the meta-governance strategy) for conflict resolution (see details in the following section).

Government strategies for conflict resolution

As argued earlier, government strategies for problem solving can provide insights into government strategies for conflict resolution. Governance scholars distinguish government strategies for problem solving on the basis of different social steering approaches (Sørensen and Torfing 2007). Studies about them are mainly connected with the shift of social steering from the conventional hierarchical, top-down, or command-and-control approach to the collaboration or negotiation approach, known as *governance* (Koppenjan, Kars, and van der Voort 2009; Pierre 2000; Sørensen 2006; Sørensen and Torfing 2005; Termeer 2009). The conventional government approach implies that governments are the central rulers or regulators, and they inherently occupy superior positions over other societal actors. Governments use command-and-control for problem solving and view problem solving as a linear intellectual design. Decisions are made by political authorities, and governments implement them in a top-down way. Often, problem solving using the government approach tends to result in an adversarial relation among stakeholders (Gray 1989). The governance approach may refer to New Public Management (NPM), good governance, corporate governance, self-governing networks, the strong capacity to deliver public service, or the minimal state (see Fukuyama 2013; Rhodes 1996). Despite the fact that various meanings are attached to governance, it is mostly seen as a non-hierarchical or horizontal strategy for solving problems. Governance implies that governments do not govern above citizens, societal groups, or private groups. Rather, it stresses that governments should negotiate or bargain with other actors to collectively resolve problems (Pierre and Peters 2000; Rhodes 1997; Sørensen and Torfing 2003).

Scholars usually use the terms *alliance*, *partnership*, *collaboration*, and *governance* interchangeably. The topic of collaboration, in particular, is widely studied by scholars from different academic fields. Collaboration is defined by Gray (1989, 5) as "a process through which parties who see different aspects of a problem can constructively explore their differences and search for solutions that go beyond their own limited vision of what is possible." Different parties bargain and negotiate with one another to find common ground for achieving a jointly acceptable solution (Kickert, Klijn, and Koppenjan 1997). Collaboration is viewed as a strategy for resolving turbulent and complex problems (Axelrod 2006; Emerson, Nabatchi, and Balogh 2012; Gray 1985; Gray and Wood 1991; Huxham 2003; Huxham and Vangen 2000; Logsdon 1991a, 1991b). In addition, it is increasingly applied by governments in Western democracies for problem solving in many fields, such as the provision of public services (like healthcare, public transport, and education), the management of common natural resources, and the resolution of environmental conflicts.

Among them, the management of natural resources is one specific field of application. Some policy scholars attempt to explore how to manage common natural resources in a collaborative way (see Leach 2006; Leach, Pelkey, and Sabatier 2002; Lubell 2004a, 2004b; Ostrom 1990; Ostrom, Walker, and Gardner 1992; Weible, Sabatier, and Lubell 2004). Often, the collaboration strategy implies that the decision makers continuously interact with the other actors with the aim of achieving a mutually acceptable solution. Some scholars have found that collaboration is effective in resolving collective problems for the management of common natural resources (Imperial 2005; Lubell 2004c; Schneider et al. 2003). However, the application of the collaborative approach may result in negotiated nonsense, implying that actors agree with solutions that are ineffective and scientifically untenable (De Bruijn and Heuvelhof 2008).

The government approach (hierarchical or command-and-control) and the governance approach (negotiation, collaboration, or partnership) are not opposed to each other. Rather, they are complementary (Lubell et al. 2002). In one study, Hysing (2009a) concludes that the existing governance modes in the fields of forests and transport in Sweden have been complemented and elaborated to formulate a mix of both the government and the governance approach. His second study (2009b) further validates this conclusion.

In addition to the government and governance approaches, network management is increasingly acknowledged as a third approach to problem solving. In a multi-actor context, self-regulation, self-governance, or self-steering may fail. The issue that then arises is how to govern the interactions among actors in order to avoid conflicts and achieve collaboration. These activities are normally termed interorganizational coordination, the governance of governance, meta-governance, network steering, network governance, or network management (Jessop 1998; Klijn, Koppenjan, and Termeer 1995; Mandell 2001; Marin and Mayntz 1991; Marsh and Rhodes 1992; O'Toole 1988; Provan and Milward 1995).

Some studies, specifically by Dutch scholars, have illustrated that network management can be applied as a strategy to resolve environmental conflicts. Van Bueren, Klijn, and Koppenjan (2003) conclude that network management may have been an appropriate strategy to address conflicts such as the Dutch Zinc debate. In a case study by Huys and Koppenjan (2009), the use of network management as a way of dealing with environmental conflict regarding the extension of the Amsterdam Schiphol Airport in the Netherlands is analyzed. In the same fashion, in the book *Managing Environmental Disputes: Network Management As An Alternative*, edited by Glasbergen (1995), several scholars, mostly from Europe, analyze how network management is adopted as a strategy by governments to resolve environmental conflicts.

Specifically, three main network management strategies can be identified, namely, institutional design, process management, and network framing. They are introduced as follows.

1   *Institutional design*: Network characteristics, especially rules or institutions, formal or informal, may shape problem-solving processes (Klijn 2001; Ostrom 2011). Institutional design is defined as interventions that attempt to change the rules used in networks (Klijn and Koppenjan 2006). Van Buuren and Klijn (2006) identify three categories of strategies regarding institutional design. The first category is aimed at network composition. Some specific strategies, such as changing actors' positions, adding new actors, or advancing network formation, characterize it. The second category of strategies is aimed at network outcomes through changing actors' choices or strategies. The third category is aimed at network interactions through changing network rules.

2   *Process management*: Various actors in networks develop their strategies autonomously, and these strategies may conflict with one another. This may result in stalemates that impede the problem-solving process. Process management is then required and used mainly to improve the interactions among actors in network contexts (De Bruijn and Heuvelhof 2008). The process managers can selectively choose actors in network processes (Agranoff and McGuire 2001), shape the organizational arrangements for coordination (Koppenjan and Klijn 2004), or supervise the processes of interactions among actors (Gage and Mandell 1990).

3   *Network (re)framing*: Actors in networks may have different perceptions or objectives in relation to the nature and the resolution of their common problem (Dery 1984). Their varied perceptions may impede their interactions. Network managers may select a framing strategy to establish a common objective for all the involved actors (Sørensen and Torfing 2009) or to develop a mutual perception about the nature and the resolution of the problem (Koppenjan and Klijn 2004). The framing strategy chosen by metagovernors may take various forms, such as using administrative stories or sensitizing concepts (Klijn and Koppenjan 2006).

These three strategies for network management provide options for process managers to govern the interactions of actors in networks, especially when stalemates or deadlocks occur. However, network management is not a simple activity. Rather, it is a process of pushing and pulling among actors with uncertainties (Klijn and Koppenjan 2012). Both public and private actors can function as process managers in principle, but often public authorities play such roles (Klijn and Koppenjan 2000a; Kooiman and Jentoft 2009).

From the governance and decision-making literature, three government strategies for problem solving and conflict resolution are identified, namely, the government (command-and-control) strategy, the governance (negotiation or collaboration) strategy, and the meta-governance (network management or network governance) strategy. The definitions of government, governance, and meta-governance are presented in Table 2.2.

Explanation of why particular government strategies are applied to conflict resolution

Some governance scholars have attempted to answer the question of which factors may influence the application of a collaboration strategy for problem solving or conflict resolution. When governments realize that the traditional command-and-control strategy is costly and ineffective in resolving problems, they may adopt a collaboration approach (Imperial 2005; Weible, Sabatier, and Lubell 2004). In addition, it is argued that, if governments think that the problem is serious (Lubell et al. 2002), or if they identify common interests or coincidence in values shared with the other stakeholders (Logsdon 1991b; Roberts and Bradley 1991), or if they recognize that they are dependent on others (Gray 1985, 1989; Logsdon 1991a), they may apply a collaboration strategy. However, some conditions hinder the adoption of a governance (or collaboration) strategy, such as the lack of consciousness of interdependency with other stakeholders, institutional barriers, the primacy of politics or the political tradition, or the nature of the discourse (such as seeking firm truth for problem solving) (Blom-Hansen 1997).

In addition, public-policy and agenda-setting scholars have conducted many studies to explain policy changes. The application of government strategies is

Table 2.2 Government strategies for conflict resolution

Government strategy	Definition
Government	This strategy means that governments use the hierarchical command-and-control approach to resolve conflicts on the basis of scientific investigation and political decisions. Governments view themselves as superior to the other actors and as the center for conflict resolution.
Governance	This strategy implies that governments attempt to establish a collaborative relationship with other actors. It is characterized by interactions, negotiations, consensus-building, and bargaining among the actors involved.
Meta-governance	This strategy signifies that governments act as the meta-governors facilitating, mediating, or managing processes, setting rules, or designing or reshaping institutions to influence the processes of conflict resolution.

related to policy changes; a change of strategy is one form of policy change. Governments continuously make decisions to change or stick to their strategies or policies. Sabatier (2007) has identified three general mechanisms that trigger policy changes: (1) events-led policy change: this implies that sudden events may result in the redistribution of resources, and governments in turn have to change their existing polices (Birkland 1997; Kübler 2001; Nohrstedt 2010; Nohrstedt and Weible 2010; Thomas 1999; Weible et al. 2012); (2) learning-based policy change: this means that governments may change their belief systems, and this results in policy changes (Hall 1993; Heclo 1974; Hogan and Doyle 2007); and (3) internal negotiation-based policy change: this signifies that governments bargain and negotiate with the other stakeholders, finally causing policy changes.

Some policy scholars (such as Baumgartner and Jones 1993; Cobb and Elder 1983; Cobb, Ross, and Ross 1976; Nelson and Yackee 2012) argue that the involvement of other stakeholders who are outside formal decision-making processes may result in collective actions (such as protests or demonstrations) that these stakeholders organize to express their opposition with the aim of shaping government decisions. Collective action is thus one condition that may influence policy change. In addition, the influence of mass media (John 2006; Jones and Baumgartner 2005; Walgrave, Soroka, and Nuytemans 2007), public opinion (Jones and Baumgartner 2004; Jones, Baumgartner, and True 1998), the nature of issues (Birkland 1998; Green-Pedersen and Wilkerson 2006; Jones and Baumgartner 2005; Stone 1989), the involvement of political parties (Breunig 2006; Walgrave and Varone 2008; Walgrave, Varone, and Dumont 2006), and the involvement of policy entrepreneurs (Bomberg 2007; Kingdon 2010; Mintrom and Norman 2009; Mintrom and Vergari 1996; Nohrstedt 2011) all may exert pressures on governments, which in turn will change their policies. Finally, objective interest (referring mainly to power maximization) is a crucial consideration for governments (John 1999; Nohrstedt 2008) that conditions government priorities, based on which governments decide to change or stabilize their strategies or policies.

The public participation in planning literature

Planning is a process involving various arguments and reasoning from different actors (Fischer 2000; Forester 1989). Often, when governments decide to construct a facility, citizens may oppose this because of its undetermined risks on their health or local environment. Conflicts between governments and citizens will then follow. To prevent or resolve such conflicts, public participation is regarded as an option for governments.

The nature of conflicts in planning

Prior to the 1960s, planning was regarded as a technical field dominated by planning expertise using scientific principles (Forester 1987; Wondolleck and Yaffee

2000). Often, governments first make decisions at the beginning of a planning process, and then announce their decisions to the public (Ducsik 1981). No public participation, or only symbolic participation, occurs during such planning processes (Arnstein 1969). In the 1960s, however, this technical model was challenged because many planned projects following the formal planning procedures were rejected because of opposition from citizens. This is regarded as a conflict that emerges in the planning processes. Many scholars label this type of conflict as a NIMBY (Not-In-My-Backyard) response, which means that citizens initiate resistance to the planned trajectories and sites for facilities.

In the early 1980s, planning scholars started discussing NIMBY responses (Burningham 2000; Hunter and Leyden 1995; McAvoy 1999; Schively 2007). They are assumed to occur because people believe that they are living too near a proposed facility that may have a negative influence on them (Kraft and Clary 1991). Opposition to roads, waste incinerators, nuclear waste disposal facilities, offshore oil drilling, nuclear and other power plants, wind turbines, wind farms, and manure-processing plants is explained using the NIMBY or backyard theory (Bosley and Bosley 1988). Many planned facilities are built to cater for a general social need, but they often have negative effects at local level. Most citizens agree with the importance or necessity of these facilities in their regions, but they disagree with their construction near them. O'Hare (1977) thus views this situation as a *multi-person prisoners'* dilemma. Local residents' essential reason for opposing the construction of the facilities is that they will experience few benefits, but costs will be primarily concentrated in their community (Kraft and Clary 1991). The core of the NIMBY response lies in the fact that "costs are not borne by the people who enjoy the benefits of activities" (Wolsink 1994, 864).

NIMBY responses initiated by citizens were initially regarded as short-sighted, parochial, irrational, and costly to society. However, NIMBY as a label based on the selfishness assumption is untenable (Wolsink 2006). Two main criticisms of the NIMBY label as explaining resistance to the construction of facilities are identified and elaborated as follows.

1   *The NIMBY response theory is unconvincing in explaining citizens' opposition to the construction of facilities.* Kraft and Clary (1991) have empirically illustrated that the NIMBY reasoning fails to characterize the majority of the testifying statements made by local residents who oppose the construction of radioactive waste disposal facilities in the US. Many testifying statements are not egoist and emotive. Opponents rationally express their worries about the construction of radioactive waste disposal facilities. In the same fashion, Wolsink's (2000) analysis, based on a large-scale survey, has also illustrated that NIMBY preferences explain only 4 percent of the variance in local residents' opposition to wind power facilities in the Netherlands. Rather, opponents' perceptions of the fairness and equity of the decision-making processes are crucial determinants of their resistance (Wolsink and Devilee 2009). Citizens' opposition to hazardous waste facilities thus cannot be simply explained in terms of self-interest, and the NIMBY theory is an

out-of-date explanation of project opposition (Hunter and Leyden 1995). It has even been claimed that the term NIMBY should be abandoned (Burningham 2000).

2   *Labeling resistance as a NIMBY response is regarded as a strategy applied by governments to disqualify the opposition of local residents.* The term NIMBY is mostly regarded as pejorative. Its main implication is that everyone prefers not to have facilities in their own backyard because they are attempting to protect their own interests. However, it is misleading to criticize local residents' rationalization as being entirely self-interested (McClymont and O'Hare 2008). In fact, the motivations behind the opposition initiated by opponents are complicated and varied (Burningham, 2000). Some environmental NGOs, for example, are opposed to the construction of facilities because of their potentially negative influence on the local environment, not just because these facilities will be constructed near their communities and endanger their interests (McAvoy 1998). In addition, some citizens oppose the construction of facilities anywhere, not just near them (Heiman, 1990). Often, when governments or planners frame all citizen-initiated opposition to facilities as NIMBY responses, they implicitly argue that the planning processes for the construction of facilities should rely on bureaucratic decision making rather than democratic processes involving public participation. Thus, the use of the NIMBY argument is regarded as a strategy by planners or governments to disqualify their opponents by implying that local citizens are selfish and have no regard for the interests of the whole community (McClymont and O'Hare 2008).

These two criticisms challenge the use of the easy NIMBY label as an appropriate explanation for local residents' resistance to the construction of facilities near their communities. This has two general implications for the understanding of the nature of environmental conflicts. One implication is that it is not always right to assume that opposition during environmental conflicts can simply be analyzed as being based on rational hypotheses. Many different actors with varied motivations or perceptions are involved in opposition to facilities. More empirical study is thus warranted to further explore this (Wolsink and Devilee 2009). Furthermore, the NIMBY response is part of a wicked or intractable problem (Fischer 1993). There are no clear-cut criteria to judge the solution to wicked problems, and this makes the intellectual design in planning processes ineffective in coping with NIMBY responses. Public participation is regarded as a prescriptive approach to resolve NIMBY responses initiated by citizens (McAvoy 1998).

Government strategies for public participation in planning

As early as 1969, Sherry Arnstein had developed an influential typology about public participation in planning. It is increasingly acknowledged that, although public participation is something that governments are likely to view as theoretically beneficial, its width (primarily the degree to which the involved actors are offered the chance to influence decision-making processes) and depth (the affected stakeholders' influence on decision-making processes and outcomes) in practice are varied (Fung 2006). Many specific strategies with regard to public participation in planning can be identified, such as manipulation, education (lecturing), therapy, defense, informing (notification), consultation, and advice (expression) (Burke 1968; Connor 1988; Ducsik 1981; Edelenbos and Klijn 2006; Plummer and Taylor 2004). Generally, these strategies match the top-down style of planning, in which planners or governments use the hierarchical approach to make and implement decisions. To oppose government decisions made using this hierarchical approach, the other actors may choose protest actions to block or redirect governments' decisions (King, Feltey, and Susel 1998). If governments choose these specific strategies, the result tends to be zero-sum or winner-take-all outcomes (Forester 1989).

The above strategies, however, do not map the whole picture of public participation in planning practice. Some authors have pointed out that these strategies are forms of what is variously referred to as symbolic public participation (Arnstein 1969), lower level of public participation (Edelenbos and Klijn 2006), or conventional participation (King, Feltey, and Susel 1998). These imply that the affected stakeholders have only a token influence on government decisions during the planning processes, and they have little power to induce the planners or governments to take their views, aspirations, or needs seriously (Fischer 1993).

Because planning proposals initiated by governments in Western democracies have met substantial opposition from citizens, planning practice has increasingly witnessed a shift toward a more open and collaborative style (Altshuler 1965; Forester 1999; Frame, Gunton, and Day 2004; Innes 1996; Innes and Booher 1999a, 1999b). Some scholars view the planning process as open when all the affected actors can play crucial roles in the production of the final decisions. The key assumption is that responsibility for planning should be directly delegated to stakeholders who work together in face-to-face negotiations to reach a consensus (Booher and Innes 2002; Wondolleck and Yaffee 2000). Some scholars view planning as collaborative: all the affected stakeholders are involved in the whole decision-making process in order to jointly find solutions (Innes 2004). Specific to the location issue for the construction of facilities, Ducsik (1981) argues that collaborative planning means that all the affected actors should first be involved in the development of the methodology and criteria to guide the process of choosing sites, and then establish an inventory of acceptable sites in an open way. Collaborative planning can result in some advantages, such as the emergence of integrated solutions resulting in an all-win outcome, successful implementation because of strong commitment to the outcome, and an increase in social capital (such as trust) through public participation (Day and Gunton 2003).

It is true that the emergence of collaborative or open planning has many implications for public participation during the planning process. Fischer (1993), for instance, views collaborative planning practice as essentially participatory because governments (or experts) involve the affected stakeholders in the hope of bringing the community's local knowledge to their decision-making processes. He further concludes that participatory planning can avoid the zero-sum politics of NIMBY, which is characterized as conflictual interactions. Innes and Booher (2004) regard public participation in collaborative planning as *collaborative participation*, a new paradigm in participation, which implies a multi-way interaction in which governments and other stakeholders debate with one another in formal and informal ways to influence the decision-making process. In addition, collaborative participation means that the interests of all stakeholders are addressed during the planning process, and they are treated equally during the discussions (Innes and Booher 2004). Some terms, such as co-producing or co-deciding (Edelenbos and Klijn 2006), genuine participation (Arnstein 1969), co-designing (Enserink and Monnikhof 2003), and authentic participation (King, Feltey, and Susel 1998; Roberts 1997) characterize the nature of public participation in this new planning practice.

Arnstein (1969) proposes that, when citizens have real power during planning processes, genuine participation will emerge. Two forms of genuine participation are identified. (1) Partnership: this implies that the affected stakeholders are genuinely involved in the planning process, and they negotiate and bargain with governments to make decisions. This form of participation corresponds with Edelenbos and Klijn's (2006) co-producing of decisions in planning processes, implying that governments and other involved stakeholders jointly establish a problem-solving agenda and search for solutions. (2) Citizen control: citizens are fully in charge of policy and managerial aspects. This can also be viewed as co-deciding, in which governments leave the decision-making process to the involved actors, and they play mainly advisory roles. As noted in the opening paragraph of this section, genuine participation relates not only to the depth of public participation (mainly the affected stakeholders' influence on decision-making processes and outcomes), but also to the width of public participation (primarily the degree to which the involved actors are offered the chance to influence decision-making processes). Enserink and Monnikhof (2003) propose a co-designing planning process: the affected stakeholders are involved in the planning processes at a very early stage, from problem recognition and definition to environmental impact assessment.

In the same fashion, the term *authentic participation* has virtually the same meaning as genuine participation. It refers to a meaningful participatory process, and its key elements are commitment, trust, and open and honest discussion. Citizens are central in decision-making processes, and they have an immediate and equal opportunity to influence them (King, Feltey, and Susel 1998). Booher and Innes (2002) regard this as authentic dialogue; this means that all the stakeholders can speak with one another with sincerity, accuracy, comprehensibility, and legitimacy with the aim of achieving joint fact-seeking. Some scholars view planning as a practical process of argumentation, implying that different stakeholders involved in planning practices argue with one another to influence the discourse (Fischer and Gottweis 2012). In addition, authentic participation can be a deliberation process, in which the involved stakeholders can debate with one another and learn from one another, and finally find a common understanding about how to resolve a problem (Roberts 1997). Authentic participation may be ideal and will never be fully achieved (Innes and Booher 2004). However, it provides some promise that can be advanced in practice.

In summary, on the basis of the degree of public participation, the literature on public participation in planning discusses two essentially different public participation patterns: symbolic participation and authentic participation. In planning practices, governments can select strategies based on the degree of participation, and two general government strategies can be identified: symbolic participation and authentic participation. They are presented in Table 2.3.

Explanation of why specific government strategies are applied during planning processes

Some scholars have argued that it is necessary to transform representative into participatory democracy (Dryzek 1990). Participation is regarded as *taken-for-granted*, implying a shared but unspoken assumption of governments in Western democracies (Majone 1992). A set of concepts, methods, and instruments for the running of participatory democracies has been devised by scholars (Ackerman 2004; Fung 2005; Fung and Wright 2003). Governments normatively view public participation as an appropriate action.

Sometimes, politicians and government officials engage citizens in participation for political expediency and power maintenance (Hoppe 2011a). In practice, governments have a menu of participation options from which they can choose. Although higher-level governments may mandate *maximum feasible participation*, local governments may change this to *advisable citizen participation* (Roberts 2004). Bishop and Davis (2002, 21) conclude that government choices of participation options are "shaped by the policy problems at hand, the techniques and resources, and, ultimately, a political judgment about the importance of the issue and the need for public involvement." This implies that governments choose their strategies regarding public participation on the basis of the *logic of consequence*.

Table 2.3 Government strategies for public participation in the planning process

Strategy	Definition
Symbolic participation	Governments dominate the planning processes and the public fail to actually influence them. Public participation is superficial and meaningless.
Authentic participation	Planning is an open process, and all the affected actors have the chance to become involved in it. Decisions during the planning process are made collectively by all the stakeholders.

Few studies have been done to specifically explain governments' choice of strategies in relation to public participation in planning processes. However, the two general explanations identified above, the rational (or the perceived benefit-cost) and the norm-based explanation, can shed light on this issue. It can be concluded that normative value and the interests of governments influence their choice of strategies to apply to public participation in planning processes.

The social conflict resolution literature in China

Citizens in China have few institutionalized channels to be directly involved in formal decision-making processes. Sometimes, they adopt contentious ways, such as mass demonstrations or violent confrontations, to express their grievances. These in turn may result in conflicts between citizens and governments. These conflicts are regarded as social conflicts in the Chinese context, and Chinese governments adopt varied strategies to address them.

The nature of social conflicts in China

Social conflict emerges when different social groups clash over antagonistic interests (Yu 2007). The Chinese state establishes "an overriding need for stability" and "harmonious society" as its official policy line and views social conflict as a taboo (Yu 2007, 2). However, since China's opening up in 1978, the prevalence of social conflicts around China has turned it into a contentious society (Chen 2012). Social conflicts arise in many fields in China, such as *hukou*,[1] tax, land appropriation, immigration, and healthcare, among others (Perry and Selden 2010).

Yu (2007) distinguishes two stages in social conflicts in China. The first stage arose between intellectual elites and political elites who struggled for political power. The 1989 Tiananmen Incident signified the end of social conflict in this stage. The second stage of social conflict emerged at the end of the twentieth century. Governments, business professionals, and academia formed an elitist alliance that enjoyed the benefits of China's economic achievements. However, the interests of poor and disadvantaged citizens, especially peasants and workers, were ignored. As a result, the latter organized various contentious actions to protect their rights.

Some scholars study social conflicts in rural areas of China (Deng and Yang 2013; O'Brien and Deng 2015; O'Brien and Li 2006). *Rightful resistance*, a concept originally applied in studies on farmers' protest actions in rural China, was initially devised by O'Brien (1996). It is primarily used to characterize situations in which the policy-based resisters cite laws, government documents, or even political propaganda to defend their legitimate rights and interests (O'Brien and Li 2006). In general, rightful resistance is a form of popular contention characterized by the following features: (1) it operates near the boundary of an authorized or institutionalized formal channel, (2) it uses the rhetoric and commitments of the powerful to limit political or economic power, and (3) it hinges on locating and exploiting divisions among the powerful (O'Brien 1996). Rightful resistance is regarded as a boundary-spanning contention, and rights-conscious citizens tend to utilize *rights upholding (Weiquan)* language to call on government officials to implement policies (Johnson 2013a). This right-based resistance has implications for understanding the nature of environmental conflicts in China. It signifies that, during environmental conflicts, Chinese citizens express their opposition to government decisions on the basis of national laws, institutions, or regulations. This may depoliticize citizens' protest actions and make it hard for governments to use state suppression.

Conflicts relating to environmental issues have been discussed by various authors (such as Johnson 2008, 2013a, 2013b, 2014; Li, Liu, and Li 2012; Li et al. 2016a, 2016b; Liu et al. 2016; Mertha 2009). These authors have studied how the construction or operation of various facilities, such as waste incinerators, dams, and chemical plants, are debated among governments, local residents, environmental NGOs, and experts.

Studies on environmental movements and environmentalism (Ho 2001; Ho and Edmonds 2007a) also provide insights into the nature of environmental conflicts in China. Some scholars attempt to answer the question of whether an environmental movement is likely in China (Tong 2005; Xie and van der Heijden 2010). It is thought that sustained contentious collective action is a requisite element of a social movement (Tarrow 1994). However, environmental movements in China do not show such a feature (Thibaut 2011), and only isolated instances of contentious politics exist (Brettell 2003). Stalley and Yang (2006) therefore conclude that nowadays China is undergoing dynamic environmentalism, but not the emergence of an environmental movement. Ho (2001) refutes the existence of an environmental movement in China too. He argues that the *greening of the state* and the ambiguous approach of the Chinese state to civil society have removed the opportunities to confront governments in China. It is thus unlikely that a confrontation-oriented environmental movement will be seen (Mol 2006). Environmentalism in China is characterized as "embedded environmentalism," implying the localized and non-confrontational nature of environmentalism (Ho and Edmonds 2007b, 333), or referred to as a "female mildness" – a greening without conflict, an environmentalism at a safe distance from direct political action (Ho 2001, 916). Derived from the concept of embedded environmentalism, three implications about the nature of environmental conflicts in China are drawn below.

1   *The depoliticized nature of environmental conflicts*: The Chinese state does not tolerate strong organized collective action initiated by Chinese citizens. When their actions are labeled as *violent anti-government collective actions*, the Chinese state will do its best to end them, even using state force. Embedded environmentalism implies that the participants involved in environmental conflicts tend more to depoliticize their actions, focusing on legal and technical issues (Johnson 2013b).

2   *The localized nature of environmental conflicts*: Western democracies have witnessed a shift from parochial NIMBY sentiments to an environmental

movement at national and global level (Rootes and Leonard 2009). The existing political structures in China greatly impede the occurrence of nationwide and well-organized environmental conflicts (Schwartz 2004). The Chinese state is extremely nervous about dispersed groups formulating a wide network that may challenge its authority (Ho and Edmonds 2007a). There is no networked environmental movement in China, and citizens tend to organize environmental protests at local level with the intention of protecting their own interests. Some issues, such as environmental quality or justice at national level, or environmental policy implementation around China, are not their concern.

3   *The non-confrontational nature (or low degree of confrontation) of environmental conflicts*: Sustained confrontation-oriented collective action is impossible in China. Some factors may explain this, such as the institutionalization of the environmental complaint system (Brettell 2003), the greening of the state (Ho 2001), the unavailability of political opportunities for collective actions (Tong 2005), the absence of urgency to confront the central government (Ho 2001), and the absence of strong middle-class support (Tang and Zhan 2008). Although environmental conflicts in China are generally non-confrontational, the sense of urgency for confrontation with governments indeed exists (Thibaut 2011). This implies that this urgency may create opportunities for the occurrence of strong confrontation-oriented collective actions during environmental conflicts.

To summarize, these above characteristics of environmental conflicts inspired by the concepts of rightful resistance and embedded environmentalism are helpful in clarifying the nature of environmental conflicts in China.

Government strategies for social conflict resolution in China

Many actor strategies can be identified in social conflicts in China. The most relevant studies are about the strategies applied by disadvantaged citizens when they confront conflicts with governments or other citizens (Cai 2010; O'Brien and Li 2006). Such citizens can adopt various strategies to resolve their grievances or complaints in relation to governments or other citizens. They can use formal channels, such as judicial or administrative channels (Goelz 2009; Michelson 2008; Warwick and Ortolano 2007). They may also apply informal strategies, such as appealing to higher-level authorities (O'Brien and Li 1999), group petition (O'Brien and Li 2006), seeking assistance from the mass media or personal networks (Shi and Cai 2006), exploring issue linkages (Deng and Yang 2013), or involving NGOs (van Rooij 2010). Tolerating the grievances can be an option for citizens when they believe that they have few possibilities to resolve their complaints (Tong 2009); or they may take illegal action for complaint expression or conflict resolution, such as protests, violent confrontations, demonstrations, or even attacks on state agencies (Cai 2010).

Besides these strategies, informal negotiation and mediation are optional strategies often applied by Chinese citizens to resolve their conflicts with governments or other citizens. Informal negotiation may be adopted as it is flexible and less costly than formal administrative or judicial approaches (Michelson 2007). During such negotiation processes, the involved disputants have face-to-face discussions about how to resolve their conflicts (Zhao 2004). Mediation is another strategy selected by citizens. Mediation is different from negotiation because it is conducted with a third party as a facilitator. It is the most preferred approach to dispute resolution for civil matters (Tong 2009). During the mediation process, the mediator or facilitator explores whether the core interests of both parties can be met and integrated, seeking common ground where mutual gains are possible (Zhao 2004).

Depending on who plays the mediator role during environmental conflicts in China, three different forms of mediation are acknowledged (Zhao 2004). The first is the so-called *people's mediation*. The People's Mediation Committee, a non-governmental organization under the leadership of local governments and local courts, conducts mediation to resolve environmental conflicts in accordance with the law. People's mediators attempt to facilitate the communication among the disputing parties and help them to achieve an acceptable solution. People's mediation is less costly and time-consuming than formal approaches to dispute resolution. The second is *administrative mediation* in which the environmental protection bureaus play the mediator role. They mediate to help the actors involved to reach an agreement on the rights and obligations of each actor. Both people's mediation and administrative mediation are forms of extra-judicial mediation. The last form of mediation is *judicial mediation*. It is a court-centered mediation strategy in which disputes among disputants are mediated by courts. In conclusion, social conflict resolution in China is highly hybrid. Citizens can apply many different strategies to achieve their goals. In addition, they tend to adopt multiple approaches concurrently to address their grievances in the hope of reaching a more desirable solution (Tong 2009).

Lastly, the literature discusses strategies that governments apply to address social conflicts, mostly those occurring between governments and citizens. First, Chinese governments may ignore or tolerate the protests initiated by citizens (Cai 2008b). They may also repress these protests by using state force (Cai 2008c), hiring thugs to attack activists (Cai 2010), punishing selected participants (Cai 2008a), isolating activists through labeling or framing (van Rooij 2010), or mobilizing informal personal networks to impede resistance (O'Brien and Deng 2015). In some cases, they may make concessions by partly meeting citizens' demands (Cai 2008b), or by organizing symbolic participation on an ad hoc basis (Li, Liu, and Li 2012). Big concessions can also be an option for governments to handle social conflicts when they meet all citizen demands (Cai 2010).

On the basis of above analysis, four government strategies to address social conflicts can be identified: tolerance, suppression, small concessions, and big concessions. They are summarized in Table 2.4.

Table 2.4 Government strategies for social conflict resolution in China

Government strategy	Definition
Tolerance	Governments are unresponsive or they ignore the other actors' complaints.
Suppression	Governments use violence or other forceful approaches to stop the actions of the other actors.
Small concessions	Governments partly meet the demands of the other actors, or organize symbolic participation.
Big concessions	Governments meet all the demands of the other actors.

Explanation of why particular government strategies are applied in resolving social conflicts in China

The literature on social conflict resolution in China reveals a lot about the explanation of the use of government strategies in social conflicts. In general, three sets of factors are identified. One set relates to social factors, such as collective action, mass media (and social media), the involvement of activists and NGOs, and casualties in collective action. The second set is primarily about political and institutional factors in China. The third is about the other factors. They are elaborated as follows.

1   *Social factors*: Protests influence the application of government strategies in Chinese social conflict resolution (Cai 2010). Protests have the potential to result in social unrest, which the Chinese state does not tolerate. Protests that are confrontation oriented and financially supported from overseas tend to result in state repression (Cai 2008a). In addition, the mass media in China influence government decisions, especially those of local governments, because they reflect the positions of the Chinese state and the Chinese Communist Party (CCP) by releasing reports and comments (Wang 2005; Yang 2005). Local governments in turn have to be prudent in applying their strategies if their actions are reported by the mass media. Moreover, the role of social media is of growing importance in shaping government decisions because they can distribute information rapidly, following which local governments may face substantial external pressure, ultimately causing changes in their strategies (Sullivan and Xie 2009). Casualties during protests may result in strong social resentment or social disorder, and the Chinese state in turn has to relieve this in order to avoid social instability (Cai 2010). This implies that the occurrence of casualties may result in changes to government strategies. Finally, the involvement of activists (such as officials within government agencies, journalists and editors, and environmental NGOs) may influence governments' decisions during the governance of environmental issues (Han 2013; Lee 2013; Mertha 2008; Xie 2011).

2   *Political factors*: Political structure and institutional arrangements matter in the application of government strategies in China. Local governments are responsible to higher-level governments, who normally do not directly intervene in social conflicts (O'Brien and Li 1999). If perceived mismanagement by local governments may endanger state legitimacy or social order, higher-level governments will intervene, and the former will change their strategies (Cai 2008b). The political opportunity structure is regarded as one important factor that influences government policies to deal with environmental debates. Xie and van der Heijden (2010) compared political opportunity structures in two cases, the Three Gorges dam (early 1990s) and the Nu River dam project (2002–2004), and found relaxed or less control by the Chinese state on the mass media and environmental NGOs. The changing political opportunity structures may provide more chances for non-state actors to participate in resolving environmental debates; this in turn shapes the strategies adopted by Chinese governments.

3   *Other factors*: Some other factors have been recognized by scholars as shaping the application of government strategies. One factor is the relationship between the actors involved in conflict resolution. It has been found that, if local governments are aligned with local industry, they are more likely to repress local citizens' actions (van Rooij 2010). The second factor is the cost of conflict resolution. Cai (2008c) concludes that, if local governments realize that meeting citizens' demands (such as the adaption of government plans) is too costly, they tend to apply avoidance or suppression strategies. The third factor is the nature of the issue (Yang 2010). Local governments in China have a great degree of discretion in policy implementation. Normally, they are tolerant of environmental education and nature conservation but intolerant of anti-nuclear protests. The fourth factor is the influence of events (Cai 2004). Both the Chinese central government and local governments are worried that Chinese citizens or international forces may use the occurrence of events (such as the party's representatives' conferences, or the government's work conferences) as opportunities to disrupt social stability. They therefore tend to take actions to avoid this while such events are occurring. The fifth factor is the role of the CCP (Zheng, De Jong, and Koppenjan 2010). The CCP controls the political careers of government officials, who thus have to take CCP positions seriously. Finally, personal networks (or *guanxi*) may also matter in shaping the strategies of Chinese local governments in formal decision-making processes (Zheng, De Jong, and Koppenjan 2010).

In all, these three sets of factors are identified as influencing the application of government strategies in Chinese governance, including the governance of environmental conflicts.

Conclusion

In this chapter, three main issues were addressed: the nature of environmental conflicts, government strategies in relation to governing environmental conflicts, and the explanation of why particular government strategies are applied. The main findings are summarized as follows.

Findings about the nature of environmental conflicts

The first issue reviewed in this chapter was the nature of environmental conflicts. Many concepts used by scholars from four strands of literature have been reviewed in this chapter.

- The environmental conflict literature provides three main points about the nature of environmental conflicts. First, environmental conflicts can be viewed as environment-induced conflicts. This implies that environmental conflicts may emerge as a consequence of environmental degradation or scarcity, and they can occur in economic, societal, religious, and territorial fields. Second, environmental conflicts derive from the allocation of natural resources, such as water, land, or oil. This definition, though rough, provides a general impression about the phenomenon of environmental conflicts. Third, the environmental conflict literature reveals a lot about the characteristics of environmental conflicts, such as their complexity, uncertainty, subjectivity, and political and risky nature. These characteristics are helpful for clarifying the nature of environmental conflicts.
- The governance, decision-making, and policy change literature reveals a lot about the nature of problems. The concepts of problem and conflict are highly interconnected – the occurrences of problems will often inherently be accompanied by conflicts. Elucidation of the nature of problems can shed light on the nature of conflicts. Environmental conflicts studied in this book involve disagreements on technical applications and/or social values. Moreover, the governance, decision-making, and policy change literature identifies the multi-actor context of environmental conflicts; this implies that all actors may interact with one another in order to jointly influence the evolution of environmental conflicts.
- The planning literature reveals that conflicts may arise during planning or implementation processes when governments' proposals to construct facilities are confronted by objections from other stakeholders. The NIMBY response is the concept used by scholars to characterize these conflicts; it assumes that citizens oppose the construction of various facilities with the aim of removing them away from their communities. The selfishness hypothesis to explain NIMBY responses is criticized. Public participation to achieve a fair, transparent, and equal planning process is viewed as an approach to addressing the NIMBY response.

•   In the Chinese context, social conflict is a taboo for the state because its occurrence may endanger social stability or even lead to an undermining of the state. Moreover, the social conflict resolution literature in China discusses some contextualized concepts that may provide insights into the nature of environmental conflicts. Rightful resistance and embedded environmentalism are two of these. Rightful resistance implies that actors involved in (environmental) conflicts tend to be careful about the boundaries of their actions, not straying too far from the institutionalized channels set by the state. Embedded environmentalism implies that environmental conflicts in China are mostly de-politicalized, localized, and non-confrontational.

Findings about government strategies in governing environmental conflicts

Strategies to govern environmental conflicts is the second issue reviewed in this chapter. The environmental conflict literature identifies three government strategies for environmental conflict resolution, namely, the traditional, the non-assisted negotiation, and the assisted negotiation strategy. The governance, decision-making, and policy change literature reveals that governments have three general options for governing conflicts: government, governance, and meta-governance strategies. Two generic government strategies regarding public participation are identified in the planning literature: the symbolic participation and the authentic participation strategy. The social conflict resolution literature in China recognizes four government strategies for social conflict resolution: tolerance, suppression, small concessions, and big concessions.

In general, the three categorizations from the international literature all acknowledge the transition from a traditional command-and-control approach to a negotiation- and bargaining-based approach in conflict resolution. Although the social conflict resolution literature in China identifies some contextualized government strategies applied by Chinese governments, it fails to recognize governance and meta-governance as potential government strategies for governing conflicts. It may be worthwhile to integrate the government strategies derived from the international literature and the Chinese literature and establish a typology that can be specifically used to analyze government strategies during Chinese environmental conflicts. In Chapter 3, this typology is established and articulated in detail.

Findings about the explanation of why particular government strategies are applied in environmental conflicts

The third issue reviewed in this chapter is the explanation of the application of government strategies in environmental conflicts. This issue is a bit underdeveloped. The main findings are presented as follows.

•   The literature on environmental conflicts assumes that the high costs of, and dissatisfaction with, the traditional judicial approach facilitate the application of a negotiation strategy by the disputants in environmental conflicts.

- Governance scholars acknowledge both the facilitating and hindering conditions for the application of the governance (or collaboration) strategy in problem solving. Some facilitating conditions are: awareness of interdependence with other actors, the seriousness of the problem, collaborative advantages, and the coincidence of interests among actors. Some hindering conditions are: lack of awareness of the interdependency of actors, the primacy of politics, and the fragmented nature of the political structure.
- Policy scholars recognize a set of factors that may influence policy change and stability, such as events, learning, interests, mass media involvement, the influence of outside groups, policy entrepreneurs, the nature of the problem, and political parties.
- The literature on public participation in planning reveals that the logics of both appropriateness and consequence matter in explaining the application of government strategies regarding public participation in planning processes.
- The literature on social conflict resolution in China reveals that a set of conditions may influence the application of government strategies for social conflict resolution, such as the intervention of higher-level governments, the involvement of national mass media or social media, casualties, personal networks, alignment of governments and local industries, protests, the occurrence of events, and the involvement of activists.

Conclusions and a next step ...

The above findings provide concepts and theoretical arguments that can be helpful in establishing a new conceptual framework to describe and explain government strategies in governing Chinese environmental conflicts. In Chapter 3, such a conceptual framework is constructed with the aim of providing an inquiry lens to explore the description and explanation of government strategies in environmental conflicts in China.

Note

1 *Hukou* is a system of household registration in China. It officially identifies a citizen as a resident of an area, including information such as name, parents, and date of birth.

3 Toward a conceptual framework to describe and explain government strategies in governing Chinese environmental conflicts

Introduction

In Chapter 2, it was shown that four strands of literature provide insights into government strategies in environmental conflicts, namely, the environmental conflict literature, the governance, decision-making, and policy change literature, the public participation in planning literature, and the social conflict resolution literature in China. This chapter is aimed at constructing a conceptual framework to specifically describe and explain government strategies in environmental conflicts in China.

A conceptual framework is defined as a rigorous means to explore the interrelationships between different concepts. It identifies the elements (mostly concepts) and general relationships among them to organize diagnostic and prescriptive inquiries (Ostrom 2007; Schlager and Weible 2013). The conceptual framework in this chapter consists of three key elements: the policy game, government strategies in environmental conflicts, and the explanatory conditions regarding the application of government strategies. The *policy game* concept is employed as a theoretical tool to describe environmental conflict processes. *Government strategy* is a concept used to categorize government actions that emerge during environmental conflicts. *Explanatory conditions* are employed in explaining the application of government strategies. As to the relationships between these three elements, propositions are first drawn in order to show how the individual conditions influence the application of government strategies. Configurational thinking is then applied to explore how the combinations of different conditions influence the application of government strategies. The three elements and the two causal reasonings together make up the conceptual framework of this book, which is used to specifically study government strategies in environmental conflicts in China.

This chapter is structured as follows. In the first section, the definition of environmental conflict, the multi-actor context of environmental conflicts, and the main elements of policy games are introduced. The next section presents six strategies applied by Chinese governments during environmental conflicts: a go-alone, a suppression, a tension reduction, a giving in, a collaboration, and a facilitation strategy. Thereafter, seven conditions that influence the application of government strategies are acknowledged: the form and scale of protest, the position of national mass media, the stage of the project, the position of higher-level governments, the involvement of activists, and the occurrence of events. In the next section, the relationships between the conditions and the application of government strategies are articulated in detail. Seven propositions are first drafted, presenting the relationships between the individual conditions and the application of government strategies. Then configurational thinking is introduced as a way of exploring how the combinations of various conditions explain the application of government strategies. In the final section, the conceptual framework constructed in this book is summarized.

Policy game as a concept to conceptualize environmental conflict processes in China

This section discusses the nature of environmental conflicts with the aim of clarifying the phenomena studied in this book. Environmental conflicts are assumed to occur in a network context, and their resolution is featured as a game-like process. This section is structured as follows. In the first subsection, the definition of environmental conflict is presented. The definition of policy game and its key elements are elaborated in the following subsection and in the final subsection, some key arguments are made to legitimize the application of the network perspective to analyze environmental conflicts in China.

The definition of environmental conflicts

In Chapter 2, derived from four strands of literature, different understandings of environmental conflicts were identified. Some authors view environmental conflicts as environmentally induced, mostly related to environmental degradation or scarcity, such as wars arising from fights over oil, land, water, or gas, and conflicts in multi-ethnic or multi-cultural societies as a result of environmentally caused migrations (see Homer-Dixon 1999). Some authors regard environmental conflicts as debates concerning the allocation of natural resources (see Glavovic, Dukes, and Lynott 1997). These two definitions of environmental conflicts are too broad and therefore unhelpful for clarifying the scope of environmental conflicts. In addition, environmental conflict can also be seen as a NIMBY response to planning; this means that local citizens oppose the construction of facilities in order to move them away from their communities. Although researchers criticize the unscientific nature of the assumption of selfishness to explain the NIMBY response (see Burningham 2000; Hunter and Leyden 1995; Kraft and Clary 1991; Schively 2007; Wolsink 1994; Wolsink and Devilee 2009), the phenomenon that the NIMBY response describes is useful for specifying the phenomenon of interest in this book. Environmental conflicts are defined as disagreements among various actors (such as local residents, non-governmental organizations, key government actors, experts, and business representatives) concerning the planning, construction, and operation of industrial plants in urban

China. Three viewpoints are proposed in the following in order to further specify the characteristics of the environmental conflicts studied.

First, the environmental conflicts studied in this book are mostly site-specific (Bingham 1986). Site-specific environmental conflicts are typically related to the allocation of particular natural resources, locations, or situations. This book mainly concerns debates regarding the location of waste incineration power plants and paraxylene (PX) projects.

Second, the environmental conflicts studied are problems inherently involving disagreements on scientific knowledge or social values (Hoppe 2011a). The two types of conflicts studied, regarding waste incineration power plants and PX plants, involve high levels of disagreements on social values and the application of technology. Application of the technologies for producing PX and for waste incineration involves high uncertainties or risks to the local environment and citizens' personal health. In turn, citizens have a low level of acceptance regarding the application of these technologies.

Third, the environmental conflicts studied in this book are one type of social conflict in China (Yu 2007). In the Chinese context, social conflicts are mostly characterized as confrontational protests initiated by disadvantaged citizens to protect their own interests. They may occur in many fields such as land disputes, tax, or immigration. The environmental conflicts studied in this book represent disagreements among various actors (such as citizens, governments, experts, activists, and NGOs) regarding the planning, construction, and operation of industrial facilities in urban China.

Policy game as a tool to analyze environmental conflicts in China

Two perspectives on networks were discussed in Chapter 2. The first views networks as an institutional context in which conflicts occur (Klijn and Koppenjan 2000b), and the second regards networks as a normative steering approach for conflict resolution, mostly referred to as *governance* or *collaboration* (Kooiman 2003). Networks in this book are defined as an institutional context or a multi-actor context in which environmental conflicts occur. In addition, the second perspective on networks is used because governance or collaboration is established as a government strategy for environmental conflict resolution, mostly compared with the government and the meta-governance strategy (see Chapter 2).

From the network perspective, conflict resolution, including environmental conflict resolution, is not a rational linear process, but a game involving a sequence of interactions among multiple actors (Allison 1971; Crozier and Friedberg 1980; Gage and Mandell 1990). The concept of *game* as a metaphor is introduced to analyze problem-solving processes and is characterized by an ever-changing set of actors and their strategies (Klijn and Teisman 1997). The resolution of environmental conflicts in this book is regarded as a policy game that occurs in a network context: diverse actors with diverging resources, perceptions, goals, and strategies, interact with one another to jointly influence the evolvement of environmental conflicts. Environmental conflicts evolve through several rounds and result in specific outcomes. To analyze the process of environmental conflicts from a network perspective, some basic elements are elaborated.

1   *Interactions and actors*: An interaction is defined as a mutually influencing relation between two or more actors (Kooiman 2003). A series of interactions occurs during environmental conflict resolution. Interactions in networks influence the evolvement of environmental conflicts. An interaction always involves actors. A variety of actors may be involved in environmental conflicts. Individuals, groups, and organizations from the public, semi-public, and private sectors are potential actors (Klijn and Koppenjan 2016).

2   *Resources and interdependence*: Environmental conflict resolution requires a variety of resources owned by various actors. Actors' resources can be formal (such as formal decision-making powers, human resources, and formal competencies) or informal (such as legitimacy, strategic capability, and mobilization power) (Scharpf 1978). In addition, they are mostly not possessed by any single actor. Rather, they are dispersed among actors. As a result, actors depend on one another to achieve the resolution of environmental conflicts. The interdependence relationship among actors is the core condition facilitating the formation of networks (Agranoff and McGuire 2004; Marin and Mayntz 1991). Moreover, the degree of dependency varies because resources have diverse importance to the various actors.

3   *Perceptions and objectives*: Perceptions are the perceived realities of the involved actors (Koppenjan and Klijn 2004). They are the mediators between complex realities and behavior. Actors use the images built on their perceptions to choose their objectives and strategies during environmental conflicts (Schön and Rein 1994). Objectives during policy games are concrete transitions of (some) perceptions. Different objectives can be established based on one perception. To take the smog issue in China as an example, many actors perceive it as an environmental problem. Nevertheless, they establish different objectives, such as reducing industrial emissions, removing the polluting plants to other locations, or improving environmental consciousness.

4   *Arena*: Environmental conflict resolution occurs in arenas. An arena is a place or field in which actors interact with one another (Ostrom, Walker, and Gardner 1992). In reality, arenas can be management teams, project groups, or formal and informal meetings in which policies and policy measures are made and implemented. Often, the problem owners, the solution owners, and the decision makers are located in different arenas (Kingdon 2010). This implies that environmental conflict resolution is most likely when these actors converge with one another.

5   *Process*: Environmental conflict resolution is not a linear process. Rather, it is an erratic game involving several rounds (Crozier and Friedberg 1980;

Radford 1977; Teisman 2000). A round is a dynamic process in a policy game. It may begin with one actor's initiative that triggers the process of a round. Conflicts may follow. Actors will then explore opportunities to achieve jointly acceptable solutions to their common problems. However, impasses, stalemates, or stagnations may happen. A crucial decision is finally made, providing a jointly agreeable solution to the common problem. A crucial decision is regarded as a solution to the question that is central in a particular policy round (van Bueren, Klijn, and Koppenjan 2003). Following a crucial decision, the next round of policy games will begin. A crucial decision thus is viewed as a boundary to differentiate various rounds in environmental conflicts. An environmental conflict may include several crucial decisions.

6   *Strategy*: In environmental conflicts, actors with diverging perceptions and resources apply various strategies to achieve their objectives. The *strategy* concept is used to categorize the concrete government actions that emerge in environmental conflicts. Strategy is defined as a pattern formed in a set of decisions (Mintzberg 1978). This means that, if a set of decisions shows consistency over time, a decision pattern can be identified. Then, a strategy can be acknowledged. Different decision patterns identified in environmental conflicts signify various strategies.

7   *Outcomes*: Outcomes in environmental conflicts can be generally categorized into three groups: substantive outcomes, process outcomes, and institutional outcomes (Klijn and Koppenjan 2016). Substantive outcomes relate mainly to the question of whether an environmental conflict is resolved or not. They can be non-decision making, a one-sided outcome, or a win–win outcome (Klijn and Koppenjan 2000b). Non-decision making implies that environmental conflict is unresolved, and its solution is not achieved. This is an all-lose outcome. A one-sided outcome means that only a small number of actors are satisfied with the result. This is a win–lose outcome. A win–win outcome implies that all the involved actors increase their benefits. As regards outcomes at process level, these may relate to the duration of conflicts, interactions costs, fairness, transparency, and so forth. The involved actors intentionally adjust their strategies to achieve their objectives or a jointly acceptable outcome. As regards outcomes at institutional level, these are primarily about the creation of enduring relations and the formulation of the perception of mutual language or trust. Some policy changes or institutional changes occur, for instance, to expedite the processes of environmental conflict resolution. They characterize outcomes at institutional level.

These elements are used to describe the game-like process of environmental conflicts in China.

Conclusion: understanding environmental conflicts in China from a governance network perspective

Inspired by governance network theory, we acknowledge that environmental conflicts occur in a multi-actor context (Klijn and Koppenjan 2000b). This perspective helps to conceptualize environmental conflicts in China. Nowadays, Chinese governments increasingly realize the weaknesses of the top-down approach for resolving environmental conflicts (Johnson 2013b), and a large number of environmental NGOs and environmental activists participate in the resolution of such conflicts (Yang 2005). In addition, many institutions have been designed by the Chinese central government with the aim of facilitating public participation to resolve these conflicts (Johnson 2014). More importantly, interactions among governments and other actors, such as citizens, professional experts, activists, media reporters, businessmen, and environmental NGOs, have become possible, and this may shape the processes of environmental conflict resolution (Johnson 2014). Based on these arguments, it can be said that environmental conflicts in China occur in a multi-actor (or network) context. Different actors (such as local governments, citizens, experts, environmental NGOs, and media reporters) interact with one another in order to jointly influence the evolvement of environmental conflicts.

Six government strategies in Chinese environmental conflicts

This study is about government strategies in environmental conflicts. From a hierarchical perspective, government in China can be classified as central government and local governments (including provincial, municipal, district, county, and township government). It can also be classified in a sectorial way, and different government agencies can be identified, such as the city planning bureau, the environmental protection bureau, the development and reform commission, the land management bureau, the city management bureau, and the other government sectors. Local governments in China assume the main responsibility for governing local affairs, including environmental conflicts. This book focuses on municipal and district governments.

In Chapter 2, four categorizations of government strategies derived from four strands of literature were identified. They are briefly reintroduced as follows.

1 Three government strategies are identified in the environmental conflict literature to govern environmental conflicts: the traditional, the non-assisted negotiation, and the assisted negotiation strategy (Amy 1983; Bingham 1986; Susskind and Cruikshank 1987).
2 Three generic government strategies for conflict resolution are recognized in the governance, decision-making, and policy change literature: government, governance, and meta-governance (Klijn and Koppenjan 2016; Pierre 2000).
3 Two government strategies for public participation in planning processes are acknowledged in the public participation in planning literature: symbolic

participation and authentic participation (Arnstein 1969; Fischer and Forester 1993).
4   Four different government strategies to address social conflicts are recognized in the social conflict resolution literature in China: tolerance, suppression, small concessions, and big concessions (Cai 2010).

The governance, decision-making, and policy change literature provides a categorization to understand government strategies for conflict resolution. It is
mostly inspired by the transition from government to governance. Such a transition implies a paradigmatic change of governance style from a hierarchical to a
horizontal approach. The two categorizations acknowledged in the environmental conflict literature and in the public participation in planning literature
refer to similar developments. In short, similar developments have been identified in these three different fields of science. However, these categorizations are
all derived from governance practices in Western democracies. Their direct use
in the Chinese context may be questionable.

The categorization of government strategies for social conflict resolution
recognized in the social conflict resolution literature in China shows strong
similarities with these three categorizations. In addition, it throws up some speci-
fications of government strategies (such as tolerance, suppression, or concession)
in the Chinese context. However, this categorization views social conflict resolution in China as a zero-sum game or a suppression–concession dilemma. It
fails to recognize that social conflicts can be resolved through a consensus-based
approach with the potential of achieving a win–win or a zero–plus outcome.

These four categorizations all fail to function as a fully fledged categorization to analyze government strategies in Chinese environmental conflicts; but
they share some similarities, and this means that they can be aligned. In this
book, I attempt to integrate them to formulate a new categorization with the aim
of identifying and analyzing government strategies in Chinese environmental
conflicts. In summary, six government strategies regarding the governance of
environmental conflicts are identified: a go-alone, a suppression, a tension
reduction, a giving in, a collaboration, and a facilitation strategy (Li 2017).
This categorization of government strategies used in this book and its relationship with the other four are presented in Table 3.1. In the following, the six
government strategies for governing environmental conflicts are introduced in
detail.

1   *Go-alone strategy*: This strategy means that local governments have a strong
commitment to achieving their own objectives, and they have no intention
of making concessions or finding alternatives. This strategy partly matches
the tolerance strategy selected by the Chinese state to handle social conflicts
(Cai 2010). The tolerance strategy identified by Cai (2008c), however, is
only one indicator of the go-alone strategy. Other indicators, such as
defense, ignoring, avoiding, explaining, educating, lecturing, persuading, or
framing, all characterize the go-alone strategy.

Table 3.1 Overview of the categorization of government strategies regarding environmental conflict resolution

Literature/Strategy	Environmental conflict	Governance, decision making, and policy change	Public participation in planning	Social conflict resolution in China	Government strategy in this book
1	Traditional	Government	Symbolic participation	Tolerance	Go-alone
2				Suppression	Suppression
3				Small concessions	Tension reduction
4				Big concessions	Giving in
5	Non-assisted negotiation	Governance	Authentic participation	–	Collaboration
6	Assisted negotiation	Meta-governance		–	Facilitation

2   *Suppression strategy*: This strategy means that local governments have a zero or little tolerance of the existence of contrasting opinions or viewpoints during environmental conflicts. It partially corresponds with Cai's (2010) study on government strategy for social conflict resolution in China. The suppression strategy implies the punishment of all participants when they express their complaints in a collective way. It refers to government actions, such as state suppression, information blocking, isolation, or coercion to keep actors away from protests that may threaten the authority or legitimacy of governments.

3   *Tension reduction strategy*: This strategy means that local governments concede some unimportant aspects or in a late stage during environmental conflicts. They engage in actions in the hope of relieving the anger or disappointment of the other actors. This strategy is consistent with the small concession strategy proposed by Cai (2010), which means that governments meet the demands of some actors, as well as punishing some others. The tension reduction strategy implies that governments are not yet the losers, and they may still be the winners as their objectives may be achieved later. It is characterized by concrete government actions, such as symbolic dialogue with local residents or the temporary halting of the debated projects with the intention of calming the other actors down.

4   *Giving in strategy*: This strategy implies that governments radically give up their initial objectives, and they fail to accomplish their original goals. It has a same meaning as the big concessions strategy proposed by Cai (2008c), which means that governments meet all the demands of the other actors. Specific actions, such as the unilateral relocation or cancellation of the debated projects, characterize this strategy.

5   *Collaboration strategy*: This strategy equates to the governance strategy in the governance, decision-making, and policy change literature (see Ansell and Gash 2008; Klijn and Koppenjan 2016; Kooiman 2003; Sørensen and Torfing 2007), the non-assisted negotiation strategy in the environmental conflict literature (see Amy 1987; Bingham 1986; Susskind and Cruikshank 1987), and the authentic participation strategy in the public participation in planning literature (see Arnstein 1969; Fischer and Forester 1993; Innes and Booher 1999b). It means that governments have strong commitments to achieve a jointly acceptable, consensus-oriented solution. This strategy is likely to result in a win–win (zero–plus or mutually beneficial) solution. Specific government actions, such as consultation to seek consensus, compensation for a win–win solution, or information disclosure for authentic public participation, characterize this strategy.

6   *Facilitation strategy*: This strategy corresponds with the meta-governance (or network management) strategy in the governance, decision-making, and policy change literature (Agranoff and McGuire 2001; Gage and Mandell 1990), the assisted negotiation strategy in the environmental conflict literature (Susskind and Cruikshank 1987), and the authentic participation strategy in the public participation in planning literature (Fischer and

Forester 1993). Although the involved actors have realized the importance of collaboration with the other actors to achieve a jointly acceptable solution, collaboration among them is unlikely to be advanced automatically. Blockage, stagnation, or stalemate may occur due to so-called free rider problems. Then, a facilitation strategy is needed. Governments can function as mediators or facilitators to design institutions or manage processes to facilitate or enable the advancement of the process of environmental conflict resolution. Actions, such as changes in policies or institutions, the involvement of a third party as mediator, the design of rules to enhance interactions, or the organization of public forums to facilitate the processes of environmental conflict resolution, characterize this strategy.

The above typology of six government strategies is used as a heuristic tool to categorize the concrete government actions that emerge during environmental conflicts. Their operational definitions and related indicators are shown in Table 3.2.

The conditions that explain why particular government strategies are applied in environmental conflicts

The literature review in Chapter 2 has shown that many conditions influence the application of government strategies. Some conditions identified in the four strands of literature overlap. In general, seven conditions are established as essentially key conditions. They are the form and scale of protest, the position of national mass media, the stage of the project, the position of higher-level governments, the involvement of activists, and the occurrence of events. In the following, the definitions of these conditions, as well as their influence on the application of government strategies, are articulated in detail.

Conditions 1 and 2: the form and scale of protest

In Western democracies, the role of outsider groups in shaping government decisions during decision-making processes has been acknowledged by governance and public policy scholars (see Baumgartner and Jones 1993; Birkland 1998; Sabatier and Jenkins-Smith 1993). Protest is one crucial way for outsider groups to express their opinions on government decisions. Lobbying, street actions, or mass demonstrations are possible options for them. These protests potentially damage government authority and political legitimacy. Governments in turn tend to change their strategies in order to resolve the concerns expressed by citizens.

Some scholars have recognized the influence of protests on the application of government strategies during social conflicts in China (see Cai 2008a, 2010; O'Brien and Li 2006). Local governments in China are responsible to higher-level governments, not to citizens. They do not necessarily take citizens' interests seriously. This, however, does not mean that they can ignore the occurrence of protests. Rather, they may even be more sensitive to them, to a certain degree,

Table 3.2 Overview of the definitions and indicators of the six government strategies used during environmental conflicts

Strategy	Operational definition	Indicators
Go-alone	• strong and deliberative commitment to its own objective • no intention of giving in or finding alternatives	• decide–announce–defense (DAD) • ignorance/avoidance/non-response • not informing/informing at a later stage • postponing decisions • education/persuading/lecturing
Suppression	• strong and deliberative commitment to its own objective • zero tolerance of different viewpoints and actions	• information blocking • coercion • forceful suppression • isolation
Tension reduction	• strong commitment to its own solution • making small and unimportant compromises	• temporary halting of project • symbolic compensation to bolster the original solution • symbolic participation • maintaining silence
Giving in	• making big compromises whereby governments become the losers	• project cancellation • unilateral project relocation
Collaboration	• making compromises to achieve a win–win solution or joint objective • aimed at building consensus	• information disclosure for authentic dialogue • compensation for a win–win solution • consultation to seek consensus • project relocation based on consensus among the actors involved
Facilitation	• an intention to advance the process • a third party to design rules or processes	• involving a third party as mediator • designing rules • organizing public forums to facilitate the process

Source: Revised based on Li (2017).

than elected politicians in Western democracies, because Chinese governments prioritize social stability on their political agenda, and they have little tolerance of social instability (Cai 2010).

The nature of protests initiated by Chinese citizens matters for the application of government strategies during environmental conflicts. Cai (2004) has proposed that four dimensions of protests initiated by Chinese citizens influence the application of government strategy in social conflict resolution. These are *demand* (political or not), *form* (violent or not), *intention* (intentionally confrontational or not), and *connection* (supported from overseas or not). In this book, two conditions, namely, the form and scale of protest, are included to explain the application of government strategies during environmental conflicts. They are elaborated as follows.

* *The form of protest*: This is about whether protests are violent or not. It is indicated by the presence of casualties: violent protests are characterized as protests where casualties occur. Local governments are sensitive to violent protests because they potentially endanger social order or even lead to undermining of the state. Local governments in turn will attempt to end them as soon as possible (Cai 2010). Thus it can be said that, if violent protests occur, local governments tend to apply a tension reduction or giving in strategy. If the protests are peaceful, however, local governments tend to adopt a go-alone or a suppression strategy.
* *The scale of protest*: This relates to the number of participants in protests initiated by citizens. It is different from the intensity of protests, which relates mostly to the frequency of protests. Often, local governments in China have little tolerance of large-scale protests because of their potential influence on social stability. The use of state suppression to address large-scale protests, however, is too costly for them. Local governments consequently tend to select a tension reduction or giving in strategy with the aim of ending large-scale protests. If the scale of citizen-initiated protests is small, local governments tend to adopt a go-alone or a suppression strategy, as small-scale protests are unlikely to disrupt social order.

Condition 3: the position of national mass media

Governments may selectively choose the issues to resolve as they are unable to resolve all of them. They tend to resolve the issues that receive high attention from the public and adjust their policies in order to accommodate public opinion. Some public policy scholars have recognized that mass media can influence government decisions because of their influence on shaping public opinion (Cobb and Elder 1983; Jones and Baumgartner 2005).

National mass media in China are the expansion of state power (Cai 2008b; Shi and Cai 2006). Their reports or exposure of a particular issue signify that the Chinese central government has noticed that issue. If local governments fail to respond well to the signals implied in the reports released by national mass media, they will be punished by higher-level governments, specifically by the central government. Local governments in turn tend to change their strategies to address an issue reported by national mass media.

The position of national mass media during environmental conflicts is another condition chosen in this book to explain the application of government strategies during environmental conflicts. Tight state censorship of mass media is widely implemented in China. Mass media are not allowed to freely frame social issues as they must accept the political leadership of the Chinese Communist Party (CCP) (Wang 2005). National mass media, such as China Central Television (CCTV), *People's Daily*, *Guangming Daily*, and Xinhua News Agency, are the most authoritative *month tongues*[1] of the national government or the CCP. They monopolize the official sources of news reports and assume responsibility for shaping Chinese citizens' political beliefs and faith. During environmental conflicts, national mass media may make comments about strategies applied by local governments, who view these comments as signals from the Chinese central government or the top leaders of CCP. National mass media hold two different positions in relation to the strategies adopted by local governments, namely, support or criticism.

- *Support*: This means that national mass media explicitly support the strategies applied by local governments during environmental conflicts. If so, local governments will ignore the opinions of the other actors. They therefore tend to apply a go-alone or a suppression strategy.
- *Criticism*: This signifies that national mass media publicly criticize the strategies adopted by local governments during environmental conflicts. If so, local governments must show that they are taking remedial action. They hence tend to adopt a tension reduction, a giving in, a collaboration, or a facilitation strategy.

Condition 4: the stage of the project

The cost of conflict resolution influences the application of government strategies. Governments mostly tend to adopt policies that are beneficial for their strategic objectives, such as growth, survival, economic benefits, legitimacy, or autonomy (Nohrstedt 2008). In addition, governments tend to participate in collaboration with the other actors when they identify collaborative advantages (Huxham 2003; Huxham and Vangen 2000). To sum up, cost matters in the application of government strategies.

Some scholars specializing in social conflict resolution in China have found that various costs, such as economic costs, face, legitimacy, or authority influence the application of government strategies in conflict resolution. When they believe it is too costly to resolve conflicts or to give concessions, they may refuse to respond or resort to suppression (Cai 2008b, 2008c).

For environmental conflicts concerning the planning, construction, and operation of industrial plants, the costs vary depending on the stage of the project. Two stages of contentious projects, the early stage and the late stage, imply different costs for local governments if they have to resolve environmental conflicts. Their relationships with the application of government strategies are articulated as follows.

- *The early stage*: This means the starting stage of projects, such as the planning or the initial construction stage. If environmental conflicts occur in this stage, local governments do not necessarily assume high costs on abandoning their initial plans. They thus tend to apply a tension reduction, a giving in, a collaboration, or a facilitation strategy.
- *The late stage*: This means that the debated projects have formally started construction. If environmental conflicts occur in this stage, local governments may think it is too expensive to give up their initial plans. They therefore tend to apply a go-alone or a suppression strategy.

Condition 5: the position of higher-level governments

The position of higher-level governments during environmental conflicts can explain the application of government strategies during environmental conflicts. Local government officials in China are directly appointed by the higher-level government; they are primarily responsible to it (O'Brien and Li 2006). If they fail to satisfy the demands of the higher-level government, they will be punished through political demotion, or administrative or even judicial punishments. Two generic positions, implying two conflicting values, held by higher-level governments regarding the debated projects during environmental conflicts can be identified: support and opposition. Their relationships with the application of government strategies are articulated as follows.

- *Support*: This means that the higher-level government explicitly supports the debated project. For example, it may publicly support the advancement of the debated project with the intention of achieving economic development. If so, local governments tend to apply a go-alone or a suppression strategy.
- *Opposition*: This implies that the higher-level government does not support the continuation of the debated project. Often, it demands local governments to rethink this. If so, local governments tend to adopt a tension reduction, a giving in, a collaboration, or a facilitation strategy.

Condition 6: the involvement of activists

The role that policy entrepreneurs play in policy changes has already been emphasized by some policy scholars (see Mintrom and Norman 2009; Mintrom and Vergari 1996). Policy entrepreneurs commit themselves to achieving policy changes through the exploitation of *windows of opportunity* (Kingdon 2010). Thus, the involvement of policy entrepreneurs influences government decisions.

The influence of activists on the application of government strategies in social conflict resolution has been acknowledged by some scholars (such as Cai 2010;

Mertha 2009; O'Brien and Li 2006). Activists are the coordinators, organizers, or leaders of protests, and they dedicate themselves to mobilizing and forming protests with the aim of shaping government decisions. From this perspective, activists in environmental conflicts are a type of entrepreneur. Because of the high risks of participating in protests, not all citizens in China have the confidence to be activists. However, there are some activists who dedicate themselves to changing government decisions. They strategically use existing laws, rules, policies, government regulations, mass media, or even personal networks to mobilize resources with the aim of exerting pressure on governments, who in turn may adjust their decisions.

Activist involvement can explain the application of government strategies during environmental conflicts. Two values are identified: the involvement of embedded activists and the involvement of unembedded activists. Their relationships with the application of government strategies are elaborated as follows.

- *The involvement of embedded activists*: Embedded activists mostly have a close relationship with the Chinese state, directly or indirectly. In general, embedded activists often have more opportunities than the average person to access formal decision-making processes and mobilize resources to exert external pressure on local governments. It is therefore concluded that, when embedded activists are involved in environmental conflicts, local governments tend to apply a tension reduction, a giving in, a collaboration, or a facilitation strategy.

- *The involvement of unembedded activists*: Unembedded activists mostly mean ordinary citizens who are working in private institutions or organizations in China. Normally, they have few opportunities and channels to access formal decision-making processes and possess limited resources to mobilize. They consequently have little ability to influence the decisions of local governments. It is assumed hence that the involvement of unembedded activists tends to result in a go-alone or a suppression strategy.

Condition 7: the occurrence of events

Events mostly refer to unexpected shocks, such as earthquakes, hurricanes, economic crises, wars, revolutions, nuclear power plant accidents, or shifts in the political climate. Many public policy scholars have realized that events may cause policy changes (see Birkland 1998; Cobb and Elder 1983; Nohrstedt 2005; Sabatier 2007; Weible et al. 2012). The occurrence of events is regarded as a necessary condition for policy change; if no events occur, no policy change will follow (Sabatier 1988).

In the literature on social conflict resolution in China, the role of events in the application of government strategies in social conflicts has been noticed by some scholars (see Cai 2004, 2008a). Several different patterns of events can be identified, such as natural disasters, and national or international events. Normally, their occurrence will receive substantial attention from the public and the various levels of government. Local governments thus must be prudent in applying their strategies. The occurrence of events thus influences the application of government strategies during environmental conflicts.

The occurrence of events can explain the application of government strategies during environmental conflicts. In general, two values in relation to this condition are identified: the occurrence of planned events and the occurrence of unplanned events. In the following, the two values as well as their relationships with the application of government strategies are introduced.

- *The occurrence of planned events*: Planned events refer to national events (such as the people's congress and the political consultative conference, the party's representatives' conference, or the government's work conference) or international events (such as the Olympic Games, World Expo, or other such events). Their occurrence will receive high attention from the central government. Local governments thus tend to adopt a tension reduction or a giving in strategy.
- *The occurrence of unplanned events*: Unplanned events refer to events that occur suddenly, such as earthquakes, fires, or other natural disasters. When unplanned events occur, it is difficult to judge the directions of government strategies.

To summarize, the above seven conditions are used in this book to explain the application of government strategies during environmental conflicts. In the next section, the relationships between the three key concepts, the policy game, government strategies during environmental conflicts, and their explanation, are discussed.

Relationships of key concepts

As indicated at the beginning of this chapter, a conceptual framework includes not only key elements or concepts, but also their relationships. In this section, I elaborate on how these key elements are connected, especially the assumed causal relationships between the conditions and the application of government strategies during environmental conflicts. Two reasonings are used in this book to study these relationships: propositions and configurational thinking. Propositions focus on the relationships between the individual conditions and the application of government strategies, and configurational thinking focuses on the combinations of conditions and shifts in government strategies. This section is structured as follows. In the first subsection, seven propositions are articulated, showing how the individual values of the seven conditions influence the application of government strategies. In the following subsection, the reasoning of configurational thinking is briefly introduced and inspires us to explore the relationships between combinations of values of the conditions and the application of government strategies.

Propositions regarding the application of government strategies during environmental conflicts

Seven conditions that explain the application of government strategies during environmental conflicts have been identified. The main arguments are articulated earlier in this chapter to suggest how the individual values of each condition influence the application of government strategies. Based on them, seven propositions are stated and presented as follows.

1   If violent protests occur, local governments tend to apply a tension reduction or a giving in strategy; however, if protests are peaceful, local governments tend to adopt a go-alone or a suppression strategy.

2   If protests are large scale, local governments tend to apply a tension reduction or a giving in strategy; however, if protests are small scale, they tend to adopt a go-alone or a suppression strategy.

3   If the national media criticize local governments' strategies, then local governments tend to apply a tension reduction, a giving in, a collaboration, or a facilitation strategy; if the national media support local governments' strategies, local governments tend to select a go-alone or a suppression strategy.

4   If conflicts occur in the early stage of debated projects, local governments tend to apply a tension reduction, a giving in, a collaboration, or a facilitation strategy; however, if conflicts occur in the late stage, they may adopt a go-alone or a suppression strategy.

5   When higher-level governments support the debated projects, local governments tend to apply a go-alone or a suppression strategy; however, if higher-level governments oppose the continuation of the debated projects, local governments tend to adapt a tension reduction, a giving in, a collaboration, or a facilitation strategy.

6   If embedded activists are involved in environmental conflicts, local governments tend to choose a tension reduction, a giving in, a collaboration, or a facilitation strategy; however, if unembedded activists are involved in environmental conflicts, they tend to select a go-alone or a suppression strategy.

7   If planned events occur during environmental conflicts, local governments tend to apply a tension reduction or a giving in strategy; however, if unplanned events occur, they tend to change their existing strategies, but the directions of the changes are uncertain.

These propositions are preliminary conclusions derived from existing theories and literature rather than from empirical data from fieldwork. They are not, however, used for theoretical testing as a deductive approach is (Tummers and Karsten 2012). In case studies, the propositions first help shape the initially theoretical expectations about the relationships between the values of the conditions and the application of government strategies. They give direction to the empirical studies in the next stages. Second, the propositions are reshaped and new theories are built (Eisenhardt and Graebner 2007). In this book, the reshaping of propositions can be both inductive and deductive. The inductive approach means that the original propositions are specified or rephrased. The deductive approach means that the propositions are confirmed or disconfirmed.

The propositions having been presented, configurational thinking is introduced in the following section to establish the causal relationships between combinations of the conditions and the application of government strategies.

Configurational thinking about the explanation of the application of government strategies

Configurational thinking is used in this book to explore the relationships between combinations of the values of the conditions and the application of government strategies during environmental conflicts. It sees cases as configurations of aspects or interpretable combinations of characteristics (Ragin 1987). It assumes that conditions work mostly in a conjunctural way and stresses that outcomes should be explained through combinations of values (Ragin 1987). Conditions within cases are not viewed in isolation from one another. Rather, they are interconnected. Configurational thinking emphasizes the importance of context in explaining the outcome of interest and assumes that a condition may have different impacts on an outcome because of the variety of other conditions or contexts. Take a nation state's economic level as an example. It may have different impacts on the outcome of democratic development in diverging countries because of differences in other contextual conditions, such as educational level, political system, or internationalization level. Economic level thus signifies different things for different countries. Furthermore, configurational thinking assumes that changes in one value of a condition may alter the quality of the case as a whole (Ragin 1987). In other words, configurational thinking states that "a single difference between two cases may constitute a difference in kind" (Ragin 2000, 73). For example, one case is a combination consisting of the presence of value A, value B, and value C, whereas another is a combination of the presence of value A and value B, and the absence of value C. These two cases are viewed as two qualitatively different kinds of cases.

Specifically for this book, configurational thinking reveals that the application of government strategies is explained by combinations of various values of the conditions. The use of configurational thinking enables us to explore the causal complexity regarding the explanation of the application of government strategies during environmental conflicts (discussed in Chapter 4). For example, we may conclude that it is the combination of the presence of violent protest and the late stage of the debated projects that results in the application of the suppression strategy by local governments in environmental conflicts.

To summarize, configurational thinking is especially appropriate for the analysis of combinations of conditions in explaining the application of government strategies during environmental conflicts.

Conclusion: using propositions and configurational thinking as two ways to explain the application of government strategies during environmental conflicts

In this section, propositions and configurational thinking have been introduced in order to establish the relationships between the three key concepts, specifically the relationships between the conditions for, and the application of, government strategies during environmental conflicts. Propositions are the preliminarily theoretical expectations about the causal relationships between the individual conditions and the application of government strategies during environmental conflicts; these propositions provide basic anchors for us to elucidate why particular government strategies are applied. Configurational thinking inspires us to explore the relationships between combinations of the values of the conditions and the application of government strategies. Propositions and configurational thinking are highly interconnected. Propositions on the one hand offer the directional expectations about how individual conditions influence the application of government strategies. In other words, conclusions regarding the relationships between combinations of values of the conditions and the application of government strategies can be drawn partially based on propositions. On the other hand, configurational thinking is a way of showing which conditions (or combinations of conditions) are necessary or sufficient to explain the application of government strategies. Therefore, configurational thinking specifies propositions. In conclusion, propositions make configurational thinking robust by providing directional expectations; configurational thinking further specifies propositions through revealing how combinations of conditions work.

Toward a conceptual framework to describe and analyze government strategies in environmental conflicts in China

The aim of this chapter is not to establish robust theories about the topic – government strategies in environmental conflicts. Rather, a conceptual framework was constructed in order to help organize diagnostic, analytical, and prescriptive inquiries into the description and explanation of government strategies during environmental conflicts (Ostrom 2005, 2011; Schlager and Weible 2013). Some main conclusions about the conceptual framework constructed in this chapter are drawn as follows.

First, a definition of environmental conflicts was provided in this chapter. Then it was stated that environmental conflicts occur in a network setting (or a multi-actor context) in which some institutional factors, such as rules, trust, or interaction patterns, condition the interactions of various actors. Furthermore, we propose that no single actor can resolve environmental conflicts by itself in a satisfactory way, even in an authoritarian regime like China. Environmental conflict resolution is characterized as a game-like process. Diverse actors, such as government officials, environmental NGOs, environmental professionals, activists, media reporters, and industrial enterprises, interact with one another and jointly influence the evolvement of environmental conflicts. They have diverging perceptions about the nature and resolution of environmental conflicts, and they use these to establish their objectives and adopt strategies to achieve them. Consequently, outcomes at the substantive, the procedural, and the institutional level can be identified.

Why a particular government strategy is applied in environmental conflicts is the core issue addressed in this book. The *strategy* concept is chosen to categorize concrete government actions that emerge during environmental conflicts. Six different strategies were identified that can be applied by Chinese local governments to address environmental conflicts, namely, a go-alone, a suppression, a tension reduction, a giving in, a collaboration, and a facilitation strategy. This categorization functions as a heuristic tool to describe and categorize government actions that emerge in environmental conflicts. After that, seven conditions were identified that explain the application of government strategies during environmental conflicts. These are the scale and form of protest, the position of national mass media, the position of higher-level governments, the stage of the project, the involvement of activists, and the occurrence of events.

In addition, the relationships between the conditions and the application of government strategies were established through propositions and configurational thinking. Propositions are used to show the relationships between the values of the individual conditions and the application of government strategies, and configurational thinking facilitates the exploration of the relationships between combinations of the values of the conditions and the application of government strategies. They are interconnected: propositions provide directional expectations for configurational thinking and configurational thinking is helpful for specifying propositions.

Finally, three key elements, namely, the policy game, government strategies during environmental conflicts, and the conditions that explain the application of government strategies, as well as their relationships (using propositions and configurational thinking), make up the conceptual framework used in this book. In Chapter 4, the research method and research strategies applied in this book are introduced. In Chapters 5 and 6, this conceptual framework is used to analyze two single cases to elaborately answer the research questions of how Chinese local governments cope with environmental conflicts, and to explain the application of the various strategies. In Chapter 7, a comparative case study is reported to investigate which conditions matter in explaining the application of (or the patterns of) government strategies in environmental conflicts. In Chapter 8, qualitative comparative analysis (QCA) is used to further study how combinations of the conditions shape government strategies during environmental conflicts. In Chapter 9, the main findings identified in Chapters 5, 6, 7, and 8 are summarized. Some reflections on this book are presented and a research agenda is set.

Note

1 *Month tongues* are mass media in China that release comments on behalf of Chinese governments or the CCP. Both Chinese governments and the CCP have their own mass media to express opinions, suggestions, and comments about social affairs.

4 Research strategy and research method

Introduction

This book applies a multi-method approach to studying government strategies in environmental conflicts. Three case study strategies are adopted: a single case (within-case) study, a comparative case study, and a qualitative comparative analysis (QCA). Three steps are taken in this book. Two in-depth case studies are first reported to gain in-depth knowledge about the governance of environmental conflicts in China. They help to elucidate what is going on, and this enables me to conduct a more focused and structured analysis of eight cases. Second, these eight cases are studied. Third, all 10 cases are compared (unstructured and QCA) in order to further study the governance of environmental conflicts. The comparative case study researching 10 cases of environmental conflicts allows me to study the relative importance of the individual conditions identified in Chapter 3 in explaining the application of government strategies in environmental conflicts. The comparative study using QCA suggests how combinations of the conditions influence the application of government strategies during environmental conflicts. A general overview of this multi-method research design is introduced in the first section. Thereafter, general information about the three case study strategies is introduced , and in the final section some general procedures about the combined use of the three case study strategies are detailed.

A multi-method research strategy

Since 2000, public administration studies have become more quantitatively oriented as more and more scholars prefer quantitative methods (Groeneveld et al. 2015). In addition, some scholars prescriptively propose that qualitative case studies should be conducted following the standards of quantitative methods (King, Keohane, and Verba 1994). All methods have their advantages and limitations, however (Coppedge 1999; Goertz and Mahoney 2013). In this book, I attempt primarily to address two research questions: *how* do Chinese local governments govern environmental conflicts and *why* do they apply various strategies? Consequently, the case study method is a more suitable option than a quantitative method to answer these how- and why-oriented research questions (Yin 2008).

Nowadays, there is a renewed interest in case-oriented research (Beach and Pedersen 2013; Bennett and Elman 2006a; Hall 2013). More and more scholars have started to identify the value of qualitative case studies to complement quantitative approaches by identifying the causal mechanisms to show us why the causal relationships exist and how they are generated (Brady and Collier 2010; George and Bennett 2005; Mahoney 2004). And some scholars increasingly recognize that qualitative case studies are superior in addressing causal complexities such as multiple causation (Ragin 1987), path-dependence arguments (Bennett and Elman 2006b; Pierson 2004), and two-level theories (Goertz and Mahoney 2005).

In this book, three case study strategies are used: single case study, comparative case study, and QCA. The main argument for doing so is that the individual methods have their own advantages and limitations; their combined use has the potential to provide satisfactory answers to the research questions (see Rihoux 2006; Schneider and Rohlfing 2013; Schneider and Wagemann 2012). The single case study method is first used; this is helpful for obtaining in-depth insights and knowledge about the phenomenon of interest – the application of government strategies during environmental conflicts. It provides building blocks for the research in the next stage. However, it is potentially dangerous to make inferences based on single case studies. Therefore, a comparative case study is next used; this is helpful for identifying which conditions are relatively important in explaining the application of government strategies. A comparative case study nevertheless has its limitations too. One of them is that it does not elucidate how conditions explain the application of government strategies in a conjunctural way. QCA is appropriate to explore causality through answering how different combinations of conditions, characterized as causal recipes expressed in terms of necessity and sufficiency, lead to the application of various government strategies. To sum up, the combination of three case study strategies provides the possibility to draw more robust explanations about the application of government strategies during environmental conflicts. An overview of the three case study strategies is shown in Table 4.1.

Two issues are first clarified before the combined use of the three case study strategies is introduced:

1   *The combined use of the inductive and the deductive logic of inference*: Inductive and deductive approaches are two basic ways of making inferences. The inductive inference approach is derived mainly from the grounded theory approach developed in the 1960s by Barney Glaser and Anselm Strauss, who propose that a systematic qualitative analysis can generate theories that are grounded in empirical realities (Strauss and Corbin 1990). The assumption is that patterns, themes, and categories of analysis are derived mainly from data rather than from theories. The deductive approach starts from existing theories and subsequently tests them through empirical analysis. Both the inductive and the deductive logic of inference are used in this book. Preliminary propositions derived from existing literature are first proposed using the deductive

Table 4.1 Three different research strategies in this book

Chapter	Method	Dependent variables (how government strategies are measured in various phases of this book)	Relationships between the conditions and the application of government strategy
Chapters 5 and 6	Single case studies	The individual shifts in government strategies.	How do various conditions influence the shifts in government strategies?
Chapter 7	Comparative case study	Three patterns of government strategy indicated by project continuation, project abandonment, and project relocation.	How do various conditions explain the three patterns of government strategies?
Chapter 8	QCA	Two outcomes regarding the debated projects: the occurrence of government compromise and the absence of government compromise.	How do combinations of conditions lead to the presence or the absence of government compromises?

approach in Chapter 3. Chapters 5, 6, 7, and 8 dominantly use the deductive logic of inference: the propositions are confirmed, disconfirmed, or specified. Confirmation means that empirical evidence confirms the relationships assumed in the propositions. Disconfirmation means that the relationships assumed in the propositions are not found or even falsified. Sometimes, the causality between the conditions and the outcomes of interests – in this book, the application of government strategies – may be slightly different than assumed. Then, specification is used as a qualification to refer to such situations. This is how the inductive logic of inference is used in this study. When propositions are disconfirmed or specified, a new proposition is formulated or an existing proposition is reformulated.

2   *Case selection and scope condition of this book*: The Xiamen case in 2007 was regarded as the starting point for environmental conflicts with the occurrence of mobilized protests in urban China (Ansfield 2013). The cases studied in this book cover the most intensively reported cases that occurred from 2007 to 2013. Generally, both the Chinese media and the international media reported them intensively due to the occurrence of protests. Note that the cases regarding the five PX plants are the most important ones since 2007. Environmental conflicts that occurred after 2013, such as the Yuhang waste incineration power plant case in Hangzhou in 2014, the Maoming PX case in Guangzhou in 2014, the Longgang waste incineration power plant case in Shenzhen in 2014, and the Luoding waste incineration power plant case in Guangzhou in 2015, are not studied in this book as they occurred after the data collection. In addition, some cases that occurred between 2007 and 2013, such as the PX case in Jiujiang in 2013, are not studied in this book due to a lack of empirical information. To sum up, 10 cases of environmental conflicts are studied in this book (see Appendix 1). Case studies are aimed at inductively elucidating the features of a broad population (George and Bennett 2005). It is thus necessary to establish the population or the scope condition that delimits the universe of cases for which the causal relation is to be shown (Walker and Cohen 1985). The scope condition (or population) of this book is that it concerns *environmental conflicts with the occurrence of protests concerning the planning, construction, and operation of industrial plants in urban China*. This implies that the conclusions drawn in this book can be generalized to the cases within this scope condition.

In the following three sections, the details of the three case study strategies are elaborated. In the final section, how the three case study strategies are combined is elaborated.

Single case study

Single case studies have long been criticized by scholars for selection bias, generalization, validity, and subjectivity problems (cf. Flyvbjerg 2006). In spite of these criticisms, few scholars disagree with the proposition that a single case study is an appropriate method for investigating new and complex phenomena (Blatter and Haverland 2012; Mahoney 2007). Studying strategies of Chinese local governments in environmental conflicts is a relatively new and complex topic, as the literature review in Chapter 2 has illustrated, and this qualifies the use of the single case study method at the beginning of this study.

Two single case studies are reported in Chapter 5 and Chapter 6, respectively. Some scholars (see Flyvbjerg 2006; Gerring 2007; Ragin 1992; Seawright and Gerring 2008) argue that case selection in single case studies is strategic and deliberate. Some optional strategies can be adopted, such as the typical or representative case (exemplifying a stable, cross-case relationship), the diverse case (showing the maximum variance along relevant dimensions), the extreme case (presenting an extreme value on the independent or dependent variable of interest), the deviant case (demonstrating a surprising value for a specific theory), and the influential case (choosing a case that influences the overall findings).

Researchers are allowed to choose cases according to their research questions and research aims (Flyvbjerg 2006). Selection of the 10 cases was introduced earlier in this chapter. Of these, two – the Panyu case and the Dalian case – are reported as two single in-depth case studies in Chapter 5 and Chapter 6. Two reasons legitimize reporting these two cases. The first is that I study two different types of projects, a PX plant and a waste incineration power plant. Ideally, this book should report two single case studies about two types of projects. Second, the availability of data is one important practical reason to choose the two cases. In the Panyu case in Chapter 5, both a collaboration and a facilitation strategy emerged; this is very unusual in Chinese governance. We normally think that an authoritarian state, like China, should be dominated by a top-down approach to governing social conflicts, including environmental conflicts. It is rare to see the emergence of a collaboration strategy in the governance of environmental conflicts in China. In the Dalian case, some common impressions about how the Chinese state governs social conflicts can be gleaned, such as the high alignment relationship between local industries and local governments, the priority local government places on economic development, and the dominance of the go-alone style of governance to cope with citizens' concerns or disagreements about government decisions. These characteristics fit our general expectations about Chinese governance.

Several analytical strategies can be used for single case studies, such as pattern matching (Yin 2008), the inductive approach, or process tracing (Blatter and Haverland 2012; Collier 2011; George and Bennett 2005). Among these, process tracing is one of the most popular approaches for single case studies (Hall 2013). It is a suitable approach to answer Y-centered research questions, normally characterized as the prototype of how a specific outcome (Y) is possible, implying that the researchers are interested in the many and complex causes of a specific outcome (Y) (Beach and Pedersen 2013). Process tracing is a tool to draw descriptive and causal inferences from pieces of evidence – referred to as a *temporal sequence of events* (Collier 2011). It offers a means to identify the intervening causal processes between the conditions and the outcomes (George and Bennett 2005) and gives a rich account of *how* a complex policy or decision emerges (Hall 2006; Kay and Baker 2015).

Using the process tracing approach, researchers like detectives and attorneys dig deep into cases to seek convincing and comprehensive evidence. Three different patterns of observations are necessary for its use, namely, storylines, smoking guns, and confessions (Blatter and Haverland 2012). Normally, a comprehensive explanation using process tracing should include them, based on which causal inferences can be made. *Storylines* enable us to identify the temporal proximity and succession of turning points and phases of transformation of different conditions; this provides evidence for the claim of causal connections between conditions. *Smoking guns* are evidence that provides a high level of certainty for a causal inference. The temporal and spatial contiguity between the conditions and the outcomes are the main observations that validate their causal relationship. *Confessions* provide deeper insights into the motivations, perceptions, and anticipations of main actors. These three types of observations are the main foundations from which to draw causal inferences in single case studies. In the two single in-depth case studies and in the other eight cases, I first construct the processes and relevant outcomes over time, and causal conditions are identified from document analysis. In addition, interviews with respondents, such as government officials, experts, environmental NGOs, activists, and local residents, provide additional insights into the causal relationships between the conditions and the application of government strategies. These observations enable me to make inferences about the explanation of the occurrence of, and dynamic shifts in, government strategies over time.

Comparative case study

Lieberson (1987) views comparison as the basis for all scientific approaches. He argues that social research is essentially comparative in one way or other. Campbell (1975, 180) is of the same opinion and argues that "securing scientific evidence involves making at least one comparison." However, some scholars narrow the scope of comparison. Lijphart (1971), for example, views the comparative method as a scientific method that is parallel to the statistical method, the case study method, and the experimental method. More specifically, he (1975) equates the comparative method with the comparable-case method. Comparable cases are "similar in a large number of important characteristics (variables) which one wants to treat as constants, but dissimilar as far as those variables are concerned which one wants to relate to each other" (Lijphart 1971, 687).

In this book, I follow Lijphart's (1971) understanding in viewing the comparative method as a scientific method, like the single case method, the statistical method, and the experimental method. Scholars prefer comparative case studies to single case studies because of their potential to achieve two apparently contradictory goals: in-depth understanding of the cases and generalization (Rihoux 2006). The *method of agreement* and the *method of difference* proposed by Mill

(1843) are the basis for the analytical method used in comparative case studies (Przeworski and Teune 1970). They are described as follows:

- *The method of agreement*: This method is also termed the Most Different System Design (MDSD) (Przeworski and Teune 1970). This implies that, although some cases have many differences, they surprisingly share an identical outcome. The big puzzle is how to explain this. Normally, conditions with the same value across cases are important in explaining this similarity.
- *The method of difference*: This method is also termed the Most Similar System Design (MSSD) (Lijphart 1975). Some situations or cases share many similarities. However, they have different outcomes. The puzzle again is how to explain this. Usually, conditions with different values are crucially important in explaining this.

The method of agreement and the method of difference are two different strategies. However, their differences should not be overemphasized (Przeworski and Teune 1970). Frendreis (1983) argues that both methods indeed rely on identical standards for detecting relationships and controlling for extraneous factors. They share the same line of reasoning: they both seek to identify relevant variables related to the dependent variable through covariation and eliminate irrelevant variables through the lack of covariation.

Many scholars debate the use of the two methods (see Collier and Collier 1991; Frendreis 1983; Lieberson 1991, 1994; Mahoney 2004; Savolainen 1994). The two methods are sometimes viewed as outdated and inappropriate for comparative analysis because of their failure to cope with the probabilistic perspective, data errors, multivariate analysis, and interaction effects (Lieberson 1991, 1994). Lieberson (1991) even argues that Mill (1843) as the proposer of these two methods explicitly advises against their use in social sciences as they may result in spurious conclusions.

However, some scholars argue that the two methods are useful for distinguishing potential necessary and sufficient causes: the method of agreement can be used to distinguish the necessary conditions, whereas the method of difference can be used to distinguish sufficient conditions (see Dion 1998; Mahoney 1999). In addition, the two methods can help structure comparisons and can be used to discover and confirm arguments (see Collier and Collier 1991; Skocpol 1979).

In this book, I use these two methods simultaneously, and three main arguments are articulated to justify this:

1   *The method of agreement and the method of difference are used to make different types of comparisons* (Collier and Collier 1991). The two methods are basic analytic tools to structure comparisons and identify similarities and differences in cases. The method of agreement can be used to compare cases with similar outcomes, and the method of difference can be used to compare

cases with different outcomes. In addition, their combined use allows researchers to make several different comparisons that are useful to draw robust conclusions.

2   *The combined use of the two methods makes it unnecessary to estimate the variability of the dependent variable.* Selection on the basis of dependent variables is criticized by some scholars (see Achen and Snidal 1989; Geddes 1990; King, Keohane, and Verba 1994). Their argument is that selecting cases on the basis of dependent variables entails a high probability of obtaining biased conclusions; relationships that exist between conditions and the outcomes of interest may disappear if a small sample is selected on the basis of dependent variables. However, some scholars (see Brady and Collier 2010; Dion 1998; Jervis 1989; Most and Starr 1982) contend that selecting cases on the basis of dependent variables is permissible for some types of studies, such as those with the aim of identifying necessary conditions or those using a process tracing approach. The criticism of selecting on the basis of dependent variables is not a problem if the two methods are used at the same time. This implies that we do not necessarily have to choose cases with different values on dependent variables (Frendreis 1983). Using the cases that we have, we can make different types of comparisons. The method of agreement, for example, can be applied to cases with the same dependent variable.

3   *The combined use of the two methods obviates the necessity of choosing comparable cases.* To use the two methods individually, researchers have to choose comparable cases. If we decide to use the method of agreement, we have to choose cases with many similar conditions (or independent variables), but they should have a different outcome (or dependent variable). Contrarily, if we decide to use the method of difference, we have to identify cases with many differences on independent variables (or conditions) but the same outcome (or dependent variable). The combined use of the two methods makes it unnecessary for researchers to choose comparable cases as it enables different types of comparisons (Frendreis 1983).

Because of the above advantages, the combined use of the method of agreement and the method of difference seems to be a better option compared to their individual use. In Chapter 7, both the method of agreement and the method of difference are applied to explain the application of government strategies during environmental conflicts. The method of agreement is used to identify the relatively important conditions that explain similarities in the same pattern of government strategies. The method of difference is used to identify the conditions that are important to explain differences in the patterns of government strategies. Their combined use elucidates the conditions that are crucially important in explaining the application (or patterns) of government strategies during environmental conflicts. Different from the in-depth case studies in Chapters 5 and 6, where I focus on the explanation for the emergence of individual government strategies within a specific context during environmental conflicts, the comparative case study in

Chapter 7 using the method of agreement and the method of difference allows me to look at a higher aggregation level of the 10 cases to seek an explanation for the emergence of the most dominant strategies (the patterns of government strategies) during environmental conflicts.

Qualitative comparative analysis

QCA builds on the comparative case studies initiated by Mill (1843) and Prze-worski and Teune (1970). QCA is essentially similar to the comparative case study approach, as the method of agreement and the method of difference are the cornerstones shared by both methods (Ragin 1987). However, QCA is different from it in several main aspects.

First, *QCA uses both an inductive and a deductive approach to making infer-ences* (Boswell and Brown 1999; Rihoux 2006). The use of QCA, such as the selection of conditions and their operationalization, is theoretically informed; this shows its deductive nature. Moreover, the conclusions drawn based on QCA can be elaborate and interpreted; this allows researchers to formulate new the-ories. QCA thus is inductive (Hicks 1994).

Second, *QCA integrates the advantages of qualitative and quantitative methods*. The emergence of QCA has been triggered mainly by the debates about the quantitative method versus the qualitative method. Qualitative, case-oriented, small-N, or intensive studies focus on depth, whereas quantitative, variable-oriented, large-N, or extensive studies focus on breadth (Goertz and Mahoney 2012). Ragin (1987), inspired by the idea that the two methods can be synthe-sized, created QCA to act as a methodological *third way* or a *synthetic strategy*. To sum up, QCA is both qualitative and quantitative. It is qualitative and case oriented: cases are viewed holistically as the configurations of qualitative attributes, conditions, or characteristics. It is also quantitative, because each case is reduced to a series of conditions and outcomes (variables), thereby allowing for calibration using numeric indicators.

Third, *QCA is a research approach as well as an analytical technique*. It is a research approach as it is used before and after data analysis: from a literature review, identification of the conditions and outcomes, the selection of cases, the calibration of conditions and outcomes, the construction of a truth table, the ana-lysis of the truth table, the interpretation of the analysis results, to the post-QCA analysis (Wagemann and Schneider 2010). QCA is therefore time-consuming because it involves iterative processes of comparing evidence and ideas. QCA analysis, though, is not time-consuming because it can be done through the use of the appropriate software. Furthermore, QCA has developed its own terminol-ogy. Compared to the terms *independent variable*, *dependent variable*, and *result* in regression analysis, QCA uses the terms *condition*, *outcome*, and *solution formula* or *solution term* (Schneider and Wagemann 2010).

Fourth, *set relations are the cornerstones for proposing causal claims in QCA* (Ragin 1987). In general, set relations are different from correlational relations in three ways. (1) *The set relation is about kinds or patterns of cases; correlational relationship relates to the relationship between variables.* The statement that developed countries are democratic is a set-theoretical statement, for example. It tells us the set-theoretical relation between democratic countries and developed countries. The statement that development is positively related to democracy is a correlational statement that presents the correlation between the two variables: democracy and development. (2) *Set-theoretical relations are always asymmetric whereas correlational relationships are normally symmetric* (Lieberson 1987). For correlational relationships, if the presence of an independent variable is positively related to the dependent variable, then its absence is negatively related to it. The set-theoretical relation, however, assumes that the presence and the absence of the outcome are two absolutely independent outcomes, implying that their explanations are not necessarily symmetric. The presence of one condition might be necessary for both the absence and presence of a specific outcome. (3) *The correlational relationship is insensitive to the calibrations implemented by researchers.* The correlational relation mostly calibrates its data relying on deviations from the mean. The set relation, however, is sensitive to calibration. One example is that the correlation between development and moderated development and democracy is virtually the same. For set-theoretical relations, these two sets have a different relationship to democracy, because moderated development is much more inclusive than development. This implies that the two sets have different set-theoretical relations with democracy.

Fifth, *QCA makes inferences based on necessity and sufficiency.* The formal expression of a necessary condition hypothesis is *Y only if X* (Braumoeller and Goertz 2000; Goertz and Starr 2002). Condition X is necessary if, whenever the outcome Y is present, the condition is also present (Schneider and Wagemann 2012). This implies that, if X is absent, Y cannot occur $(Y \rightarrow X)$. Conversely, when X is present, Y does not always occur. The assessment of the sufficiency of a condition needs researchers to determine whether the condition in question always produces the outcome in question (Ragin 2000). A condition can be regarded as sufficient if the outcome is present whenever it is present across cases. It can be expressed as *if X, then Y,* or *X implies Y,* or *X is a subset of Y* (Schneider and Wagemann 2012). Necessity and sufficiency are different. The former is similar to the term *precondition,* which implies that the occurrence of necessary conditions may lead to the occurrence of an outcome, and their absence will definitely be followed by the nonoccurrence of that outcome (Ledermann 2012). The latter means that the occurrence of sufficient conditions always results in the occurrence of an outcome and their absence is not always followed by the nonoccurrence of that outcome. The necessity and sufficiency of conditions are the causal relationships that QCA seeks.

Sixth, *QCA allows for the exploration of causal complexity.* One important advantage of QCA is that it enables the exploration of causal complexity, which is characterized as equafinality, heterogeneity, and conjunctural causation (Byrne 1998, 2005; Ragin 1987). *Equifinality* or *plurality of causes* implies that the same outcome can be generated through several alternative conditions (or an alternative combination of conditions). *Causal heterogeneity* rejects the conventional causal homogeneity statement that assumes that the same causal factor operates in the same way in all contexts. Rather, a specific cause has different or even opposite effects on the outcome of interest depending upon its combination with various conditions. It rejects permanent causality and emphasizes context-sensitive causality. *Conjunctural* (or *recipe*) *causation* means that conditions operate in a conjunctural way and causality is context- and conjuncture-sensitive – specific conjunctions that are temporal in time and local in place activate certain mechanisms that bring about a specific reality (Gerrits and Verweij 2013). Mill (1843) views this as *chemical causation*, implying that a phenomenon emerges consequent to interactions of appropriate preconditions. The outcome does not occur in the absence of any one ingredient.

In this book, QCA is used in Chapter 8, which explains the application of government strategies during environmental conflicts. In Chapter 7, 10 cases are compared at a higher aggregation level in order to seek an explanation for the patterns of government strategies. One limitation of the study in Chapter 7 is that it does not explore how combinations of conditions explain the application of government strategies during environmental conflicts. QCA is a very useful approach for us to remedy this limitation, allowing us to study how combinations of conditions shape government strategies (or the set-theoretical relation between conditions and the outcomes of interest) in a structured way. In the following, the QCA administrative procedures are introduced.

Step 1: Decide which type of QCA to use. Three main variants of QCA are widely used by researchers: crisp-set QCA (csQCA), fuzzy-set QCA (fsQCA), and multi-value QCA (mvQCA) (Cronqvist and Berg-Schlosser 2009; Ragin 2008; Rihoux and Ragin 2009; Schneider and Wagemann 2012).

- *Crisp-set QCA (csQCA)*: In the 1980s, the discussion on QCA was limited mostly to csQCA (Klir, St Clair, and Yuan 1997). George Boole developed an algebra suitable for analyzing variables with only two possible values, such as true (present) or false (absent). This Boolean-based analysis generally aims to address the presence or absence conditions under which a certain outcome emerges. As a language, Boolean algebra has some basic conventions: (1) an uppercase letter represents the [1] value for a given binary variable; (2) a lowercase letter represents the [0] value for a given binary variable; and (3) a dash symbol [–] represents the *do not care* value for a given binary variable, implying that it can represent either the [1] value or the [0] value. Furthermore, three basic operators are employed in csQCA, namely, logical AND, connoted by the [×] (multiplication) symbol, and logical OR, connoted by the [+] (addition) symbol. The connection symbol [→] linking conditions and the outcome expresses sufficient causal relationships, and the connection symbol [←] linking conditions and the outcome expresses necessary causal relationships.
- *Fuzzy-set QCA (fsQCA) and multi-value QCA (mvQCA)*: As many social science phenomena do not come in binary form, the dichotomization of

conditions and outcomes of csQCA was criticized (Schneider and Wagemann 2012). Ragin (2000, 2008) has developed fsQCA, which allows case membership scores to range from 0 to 1. Three qualitative anchors must be established: full set membership (1), full non-membership (0), and indifference or cross-over point (0.5). In general, crisp sets attempt to identify differences in kind, but fuzzy sets explore differences in degree. mvQCA is a direct extension variant of csQCA that enables the analysis of multi-value variables (Cronqvist and Berg-Schlosser 2009; Vink and Vliet 2013).

Decisions about which type of QCA to apply are based mainly on theoretical considerations, the quality of the empirical data, and the number of available cases (Schneider and Wagemann 2010). In this book, csQCA is employed because of the nature of the available data: data on conditions and outcomes are often secondary, limiting the possibilities of calibrating conditions and outcomes into multi-value or fuzzy sets. Also, given the number of cases in this book (10), csQCA is a better option.

Step 2: The outcome and the conditions are calibrated. After csQCA is established as the method to compare the 10 cases of environmental conflicts, all the conditions and outcomes studied should be dichotomized. This is coined *calibration*, a crucially important procedure in QCA. The calibration processes should be transparent, and the reasoning justifying the qualitative anchors should be clearly presented (Schneider and Wagemann 2012). QCA regards causality as asymmetric, implying that the analysis of the presence and the analysis of the absence of an outcome should be conducted separately.

Step 3: A truth table is constructed, showing the relationships between the values of the outcome and combinations of the conditions. After the calibration of the outcomes and the conditions, a truth table is constructed to show their values. Normally, the truth table will have 2^n rows (n refers to the number of conditions). Each row in a truth table represents a logically possible configuration of conditions. The truth table shows how the outcome is linked to each configuration and which cases cover which configuration. Some configurations are not covered by any case; these are coined *logical remainders* (Schneider and Wagemann 2012). Sometimes, contradictory rows may occur in a truth table, implying cases with the same conditions but different outcomes. Several options can be applied to resolve this: adding new conditions, recalibrating the conditions or the outcomes, or eliminating cases (Wagemann and Schneider 2010). The construction of a contradiction-free truth table is the basis for the analysis of necessity and sufficiency in the next stage.

Step 4: The necessity of the conditions is first analyzed on the basis of the truth table. It is highly recommended to analyze the necessity of conditions before analyzing their sufficiency (Ragin 2000). Consistency and coverage are two measures to assess the robustness of the necessity (as well as the sufficiency) of conditions. The consistency of a necessary condition means the degree to which the condition is in line with the statement of necessity; it assesses the degree to which the condition is a superset of the outcome. If the consistency of a condition is 1, then it is a necessary condition. However, the necessary conditions may be trivial, implying that they are present in all cases, irrespective of the value of outcomes (Braumoeller and Goertz 2000; Dion 1998). Coverage is the second indicator to assess the relevance of a necessary condition; this refers to the degree to which the outcome covers the necessary conditions. For example, air is a trivial necessary condition for the survival of human beings as it exists for both the survival and nonsurvival of human beings. If a condition has a consistency of 1 and relatively high coverage, then it is a relevant necessary condition.

Step 5: The sufficiency of the condition is analyzed on the basis of the truth table, and three solution formulas based on the strategies for dealing with logical remainders are formulated. Consistency of sufficiency is defined as "the degree to which the cases share a given combination of conditions" (Ragin 2008, 44), and coverage is "the degree to which a cause or causal combination 'accounts for' instances of an outcome" (Ragin 2008, 44). The two measures enable researchers to assess the degree to which a model explains the outcome in the cases or the relative relevance of a certain configuration of conditions (Ragin 2006). In general, it presents the proportion of the sum of the outcome membership scores that is covered by a causal condition. The analysis of sufficiency produces three solution formulas to explain the outcomes of interest: the most complex, the most parsimonious, and the intermediate solution formula. The most complex solution (or conservative) formula does not take any logical remainders into the analysis. The most parsimonious solution formula involves all the logical remainders to achieve a most simple solution formula. Regarding the intermediate solution formula, directional expectations are used. Directional expectations are theoretical hunches about the causal relationships between conditions and outcomes (Schneider and Wagemann 2012). In Chapter 8, I show these three different types of solution formulas resulting in both the occurrence and nonoccurrence of government compromises with local communities.

Step 6: The interpretation of the results. When interpreting the solution formulas, researchers should come back to the cases and link the solution formulas to empirical data in order to substantiate the solution formulas (Schneider and Wagemann 2012). Researchers are allowed to interpret the solution formulas on the basis of the iterative dialogues between the theories and empirical data (Rihoux and Ragin 2009). Many scholars agree that the complex solution formula (or conservative solution formula) is too difficult to interpret and that the parsimonious solution formula is unrealistically simple. The intermediate solution is criticized as well because it makes the distinction between theory and the analysis unclear (Baumgartner 2014).

The above steps cover all crucial procedures for conducting a QCA analysis. However, it must be noted that QCA does not always strictly follow this step-wise procedure as it is essentially an iteration-oriented method (Ragin 1987). In recent years, the number of peer-reviewed articles using QCA has increased enormously, showing its popularity (Marx, Rihoux, and Ragin 2014; Thiem and

Dusa 2013). Nevertheless, this does not mean that QCA does not have limitations. Lucas and Szatrowski (2014, 3), for example, have pointed out that

> QCA fails to find correct causal recipes, fails to replicate causal recipes across data sets that differ only owing to chance, identifies causal patterns in noncausal data, does not find the correct causal patterns in deterministic data, finds interactions even when they are absent, fails to find the correct interactions when interactions are present, selects the wrong direction of association, and finds asymmetric causation when the known causal structure is symmetric.

Some scholars have developed new methods for qualitative case studies, such as coincidence analysis (CNA) (Baumgartner 2013) and necessary condition analysis (NCA) (Dul et al. 2010). In the future, many scholars will probably continue to use QCA, and some pioneering methodologists definitely will explore new methods to further improve or replace it. Nevertheless, QCA is a great improvement in qualitative case studies in spite of its limitations (Ragin 2005). Specifically, it is a very useful method whereby to explore the causal relationships between combinations of conditions and outcomes. In this book, I use QCA to research how combinations of conditions influence the application of government strategies in Chinese environmental conflicts. Its use enables me to draw some conclusions that show how the conditions work in a conjunctural way to lead to the application of different government strategies. This is the added value of QCA compared to the explorations in Chapters 5, 6, and 7.

Toward a combination of three case study strategies

These three different case study strategies having been introduced, a general summarization is presented below.

1   *Single case studies*: Single case studies were my preferred option to obtain knowledge about the specificities and complexities of the cases under study, and they enabled me to describe and analyze the shifts in government strategies. Some in-depth understanding about the dynamic nature of the application of government strategies could be obtained. Singe case studies have limitations though. One is their generalizability. It is difficult to judge the extent to which the conclusions drawn in single case studies can be generalized to other cases. The second is that the (relative) importance of the identified conditions in explaining the application of government strategies cannot be established.
2   *Comparative case studies*: Conclusions drawn from comparative case studies are more generalizable than those derived from single case studies. Moreover, the comparative study in this book, based on the method of agreement and the method of difference, permitted the identification of the conditions that were relatively important for explaining the application of

government strategies during environmental conflicts. Finally, in contrast to QCA, the comparative case study allowed the inclusion of more details about the cases under study, and more variables and outcomes could be taken into account. However, the procedures for comparative case studies are not well structured, and the conclusions drawn from them are not robust. Moreover, comparative case study did not allow me to explore how combinations of the identified conditions influence the application of government strategies during environmental conflicts.

3   *Qualitative comparative analysis*: QCA allowed me to explore how combinations of conditions result in the application of varied government strategies during environmental conflicts. In addition, the QCA procedures are highly structured and follow standard protocols. As a result, I was able to draw conclusions with high internal and external validity. Nevertheless, QCA has limitations. Although it is a case-oriented strategy, I sometimes had to compress empirical data and information, and ignored some details in order to achieve an elegant comparison. Furthermore, some dynamic and complex causal relations regarding the explanation of the application of government strategies could not always be taken into account, such as sequence or temporality. Finally, the chosen conditions in QCA are limited. I had 10 cases, implying that I ideally should choose three or four conditions in csQCA.

Table 4.2 provides an overview of the advantages and disadvantages of the three case study strategies used.

As Schneider and Rohlfing (2013) have argued, few guidelines are available to show how different case study strategies can be applied in a combined way. It is not easy to find a way of using the three case study methods in a structured and convincing way. This book combines them in the following steps.

First, *data collection*: Empirical data for this book were collected in two ways: secondary data and primary data. Secondary data were obtained mostly from the following three sources: (1) academic articles; (2) television news reports, the internet, newspapers, magazines, forums, and blogs; and (3) government documents about formal regulations, institutions, laws, reports, and policies. Fieldwork was conducted from March to June 2014 in Beijing, Guangzhou, Nanjing, and Shanghai. In addition, 12 interviews were conducted by Dr. Yi Liu, an Associate Professor at Dalian University of Technology (DUT) between August 2011 and September 2013. Eventually, in total 32 semi-structured interviews were conducted. For details of represents see Appendix 2. Respondents included media reporters, government officials, coordinators in environmental NGOs, local residents, activists, and experts. Every interview lasted about one hour on average. Finally, over 60 pages of case descriptions and about 100 pages of interview reports were compiled. It should be noted that a large amount of data and information, from both secondary sources and interviews, was collected about four cases, namely, the Dalian PX case, the Tianjin-gwa case, the Liulitun case, and the Panyu case. For the other six cases, the data were mostly secondary.

Table 4.2 Overview of the three case study strategies in this book

Chapter	Purpose	Method	Advantages	Disadvantages
Chapters 5 and 6	Explore how local governments deal with environmental conflicts and explain government strategies within cases	Single case study	• Helps to refine the conceptual framework • Obtains in-depth knowledge about the complexities and the dynamicity of cases	The generalizability of single case studies is uncertain, and the relative importance of the conditions cannot be identified
Chapter 7	Distinguish the important conditions and identify conditions that are crucial in explaining government strategies during environmental conflicts	Comparative case study	• Conclusions are more generalizable • Identifies relatively important conditions • More variables and outcomes can be studied compared to QCA • More substantive details of cases can be shown	The procedures are not structured, and the conclusions may be not robust This method is not useful for identifying how combinations of conditions influence the outcomes
Chapter 8	Provide rigid conclusions, specifically sufficient causation, about the explanation of government strategies across cases	QCA	• Identifies necessary and sufficient conditions • Explores how combinations of conditions influence outcomes • Compares cases in a systematical and structured way	The details of the cases are not well shown, and the complexities of causal relationships are not well studied The conditions that can be studied are limited

Second, *drafting propositions*: As discussed in Chapter 3, a conceptual framework was constructed in order to specifically describe and explain the application of government strategies during environmental conflicts. Propositions deductively drawn from existing literature were drafted to show the relationships between the conditions identified in Chapter 3 and the application of government strategies. These provided some preliminary insights into the explanation of the application of government strategies. In general, they were crude and needed to be further elaborated.

Third, *applying propositions in single case studies*: In Chapters 5 and 6, two single case studies are reported; these specifically reveal how the shifts in government strategies during the Panyu case and the Dalian case can be explained. The propositions developed in Chapter 3 were used to explore the causal connections between the conditions and the application of government strategies within cases. One by-product of the two single case studies was the confirmation, disconfirmation, or specification/reformulation of the propositions.

Fourth, *applying propositions in comparative case studies*: A comparative case study is the second attempt in this book to explain the application of government strategies. The comparative case study in Chapter 7 allowed me to structure my comparisons of multiple cases and made it possible to identify the relative explanatory power of the conditions regarding the explanation of the application of government strategies. Again, the propositions developed in Chapter 3 were used to explore how the patterns of government strategies in environmental conflicts could be explained. The original propositions were confirmed or disconfirmed. Some specified propositions were drawn.

Fifth, *applying propositions in QCA*: In Chapter 8, I use QCA to explore which combinations of conditions are necessary or sufficient in explaining the application of government strategies (the occurrence and nonoccurrence of government compromises with local communities during environmental conflicts). The propositions were again used as tools to shape the directional expectations in explaining this.

In conclusion, three case study strategies are used in this book. The three methods do not do the same things. Single case studies provide in-depth insights into the dynamic shifts in government strategies over time, allowing me to explain the occurrence and individual shifts in government strategies over time. The comparative case study using the method of agreement and the method of difference enables me to gain an understanding of what explains the patterns of government strategies. I look at cases at a higher aggregation level to examine how various patterns (indicated by different outcomes in the debated projects) of government strategies are explained. This makes it possible to gain knowledge about the explanatory power of the individual conditions on the application of government strategies. QCA allows me to obtain insights into how combinations of conditions shape the occurrence of government strategies.

5 Government strategies in governing environmental conflicts

The Panyu waste incineration power plant in Guangzhou as a single case

Introduction

Guangzhou is located in southeast China, one of the most economically developed and politically open regions in the country. In September 2009, local governments in Guangzhou publicly stated that a waste incineration power plant would be constructed in Dashi, Panyu district. Afterwards, a debate, involving experts, government officials, local citizens, mass media, and activists, over where the waste incineration power plant should be constructed became a hot issue in Guangzhou. These various involved actors gave their opinions, suggestions, and viewpoints in order to shape the fate of the Panyu waste incineration power plant. Over time, local governments in Guangzhou gradually learned to open the decision-making process to involve the public in order to achieve a satisfactory outcome. One intriguing issue in this case is that local governments in Guangzhou gradually learned to facilitate, channel, and enable public participation in decision-making processes. The conceptual framework constructed in Chapter 3 is used to identify which strategies were applied by local governments in Guangzhou and to explain why they decided to do so. This chapter proceeds in five sections. The first section describes the Panyu waste incineration power plant case. The next section elaborates the strategies adopted by local governments in Guangzhou, and in the following section the application of six government strategies is explained. The case is then discussed, and in the final section key conclusions are drawn.

The Panyu waste incineration power plant case

The Panyu waste incineration power plant case is introduced in detail in this section in four parts: background, network character, process, and outcome of the case.

Background to the Panyu waste incineration power plant case

Guangzhou (widely known as Canton; less commonly known as Kwangchow) is the capital and largest city of Guangdong province in China. It is located on the

Pearl River, about 120 kilometers northwest of Hong Kong and northeast of Macau. It is one of the five national central cities in China, as well as a key national transportation hub and trading port.

In the past, the central region of Guangzhou City functioned as the political, economic, and cultural center, and its subsidiary suburbs assumed responsibility for disposing of waste produced in the city center. With the high-speed urbanization around China, however, land increasingly became a scarce resource in Guangzhou City. Subsequently, the waste disposal policy in Guangzhou was changed: each individual district in Guangzhou was responsible for disposing of the waste produced in its own administrative region.

In 2000, Panyu county was upgraded to a district of Guangzhou City. One consequence of the urbanization process in the Pearl River Delta Region was the inability to process urban waste by means of landfills. In 2002, the Panyu district government started the procedure of selecting a location for a waste incineration power plant and eventually, on August 25, 2006, Dashi was selected as the preferred location for its construction. The Guangzhou Planning Bureau and the Guangzhou Development and Reform Commission approved this decision. Subsequently, the preparatory procedure for this project's construction started. Between 2007 and 2008, the project bidding, the environmental impact assessment (EIA), and land acquisition took place.[1] On September 23, 2009, the Guangzhou Bureau of City Appearance, Environment, and Sanitation reported that, once the EIA was completed, the construction of the Panyu waste incineration power plant would formally start.[2]

The network characteristics of the Panyu waste incineration power plant case

In this case, Guangzhou Municipality was responsible for waste disposal for the whole of Guangzhou City, including the Panyu waste incineration power plant. The Panyu district government was in charge of the construction of the plant, which was to be operated by Guangri Corporation under a 25-year franchise. Guangri Corporation is a state-owned enterprise (SOE) in Guangzhou affiliated to Guangzhou Municipality. Regarding the approval procedures for the plant, Panyu district's Urban Management Bureau first submitted a proposal for its construction to Guangzhou's Development and Reform Commission. Then, the environmental impact assessment institute made an EIA. After that, the proposal would be approved by Guangzhou's Urban Planning Bureau, Guangzhou's Land Resources Bureau, and Guangzhou's Development and Reform Commission.

Although many actors – different government agencies, experts, local citizens, and mass media – were involved in this case, I categorize them into three main actors, namely, Guangzhou Municipality, Panyu district government, and affected local residents (mostly local residents in Dashi). During this case, the perceptions of the three main actors changed over time. Some general characteristics can be identified, however. Guangzhou Municipality coordinates the governance of most public affairs, including the urban waste problem. It follows

Chinese central government regulations to establish waste incineration as the main approach to disposing of waste and implements this regulation in Guangzhou. The Panyu district government assumes responsibility for resolving the waste problem in Panyu district, and it follows the regulation set by Guangzhou Municipality to resolve the waste problem through incineration (respondent 17).[3] Local citizens in Panyu, especially those in Dashi, generally disagreed with Guangzhou Municipality's decision to construct the Panyu waste incineration power plant and attempted to have it moved to another location (respondent 20).

Process: five rounds in the Panyu waste incineration power plant case

The Panyu case is divided into five rounds based on five crucial decisions. Crucial decisions are important decisions or events that result in a shift in issues being discussed, or that affect the nature of interactions. Five crucial decisions are listed as follows.

1   In September 2009, the Panyu district government publicly stated that a waste incineration power plant would be constructed in Dashi. This decision triggered debate among various actors regarding the construction of the plant. It signified the beginning of the case (or the first round).
2   On December 10, 2009, the construction of the Panyu waste incineration power plant was temporarily stopped. This decision signified a change in government decisions; this was the second crucial decision. It signified the end of the second round.
3   In November 2010, the Chinese central government ruled that local governments should provide at least three alternative locations for EIA. This was the third crucial decision as it changed the rules of the game, implying that local governments in Guangzhou had to offer at least three alternative locations before making a final decision about where the Panyu waste incineration power plant could be constructed. This crucial decision signified the end of the third round.
4   In August 2011, the Panyu district government and the Nansha district government reached an agreement that the former could construct a waste incineration power plant in any candidate location. This was a fourth crucial decision because it influenced the evolvement of the Panyu case; it meant that the Panyu district government could freely choose the location for the construction of the Panyu waste incineration power plant. This decision signaled the end of the fourth round.
5   In April 2013, the Dagang waste incineration power plant started construction, signifying the end of the fifth round.

In the following, the five rounds of the Panyu case are described in detail.

Round 1: waste incineration, the only choice?
(September–October 2009)

In September 2009, the Panyu district government stated that a waste incineration power plant would be constructed in Dashi, which is quite close to the densely populated South China Plate residential area, home to over 300,000 residents. Respondent 20 claimed that many middle-class people, such as lawyers, journalists, and government officials, lived nearby. Although Guangzhou Municipality had chosen Dashi as the location for the plant as early as August 2006, residents did not hear of the plans until September 2009. Following up on local residents' concerns, their representatives drafted a petition, titled "Strongly Opposing the Waste Incineration Power Plant Project in Dashi, Panyu District; 300,000 Residents' Life and Health Are Not a Small Problem," which was signed by residents and subsequently submitted to the Guangzhou Bureau of Environment and Sanitation. Meanwhile, local residents expressed their concerns over the planned waste incineration power plant during visits to the Panyu district government and the South China Environmental Science Institute, the body responsible for EIAs. However, the former failed to receive responses from the latter. Afterwards, some local residents put on a "Mask Show," wearing gas masks in commercial streets, wearing T-shirts, and having car stickers printed with "Opposing Waste Incineration and Protecting Green Guangzhou City." They were attempting to raise awareness of the consequences of waste incineration, as argued by respondent 17. Generally, these activities did not affect the position of Panyu district government. Some activists were summoned to the Police Department in Panyu, and they were warned to keep their distance from collective activities, as argued by respondent 17.

Later on, however, local mass media, such as *Southern Weekly*, started to report the concerns of local residents in Dashi over the Panyu waste incineration power plant, and the events were covered throughout Guangdong province. *Southern Weekly* is regarded as one of the most critical newspapers in China. Respondent 20 claimed that it functioned as an agent to deliver information between governments and local citizens in Guangzhou. When local residents near Dashi initially expressed their opposition to the construction of the plant, one counselor in the Guangdong provincial government publicly claimed that they should not oppose its construction in Dashi and that waste created in Panyu district must be disposed of in Panyu.[4] This stimulated local residents to reflect; as one of them remarked, "his [the counselor's] words are very vicious, isolating our Panyu locals and making us seem to be selfish. I now admit that incineration is an option for disposing waste."[5] Increasingly, some local residents started to reframe their augments against waste incineration. As one activist commented, "we are not against waste incineration, but against incineration of unclassified waste."[6]

On October 30, 2009, the Panyu district government organized a press conference, in which four invited experts claimed that the incineration power plant would not result in pollution or health problems, and senior government officials from the Panyu district government explained the reasons for choosing Dashi as the preferred location for the plant.[7] Immediately following the press conference, some local residents contacted mass media, such as *China News Weekly*, CCTV, and *People's Daily*; this resulted in a national debate on waste incineration. Guangzhou Municipality and the Panyu district government remained silent.

Round 2: temporary halting of the Panyu waste incineration power project (November–December 2009)

On November 9, 2009, the Guangzhou Urban Management Committee was formally established by Guangzhou Municipality in order to reconsider and reorganize waste process management practices for the whole of Guangzhou City, including waste incineration power plant construction issues. On the morning of November 23, on the Guangzhou Urban Management Committee's first public reception day, hundreds of residents assembled outside its door. Respondent 16 claimed that, when the assembled citizens realized that they were not allowed to express their complaints collectively, they visited the office of the Guangzhou Urban Management Committee one by one to express their opposition to the construction of the plant in Dashi. When they found that their complaints were not being taken seriously, the residents walked to the Guangzhou municipal building, where they clustered together and demanded cancellation of the Panyu incineration power plant project. Guangzhou Municipality responded by inviting a maximum of five representatives for a meeting, an invitation that was refused by the crowd of protesters. Respondent 16 argued that the reason was that they were afraid that Guangzhou Municipality would take revenge. After several hours of protests, the residents left. On the next day, Guangzhou Municipality stated publicly that the Panyu waste incineration power plant would not be constructed as long as it failed to pass the EIA and was opposed by most local residents. On November 25, CCTV reported the debates concerning the construction of the Panyu waste incineration power plant.[8] On December 10, 2009, the Panyu district government formally postponed the plant's construction and declared it would reconsider the plant's location, to be decided upon after a renewed discussion on the waste policy for the whole of Guangzhou, and after the Asian Games, which were planned to take place in Guangzhou in November 2010.

Round 3: organizing an expert forum (February–November 2010)

After the decision to halt the construction of the Panyu waste incineration power plant, Guangzhou Municipality organized an expert forum in order to collect opinions on urban waste management practices. In February 2010, 32 experts, both supporters and critics of waste incineration, gathered with the purpose of formulating a general principle for the processing of waste in Guangzhou City. Thirty-one experts declared that waste incineration was the preferred waste process approach, with landfill as the second preference. Respondent 17 stated that hot debates occurred during the expert forum among the invited experts.

One expert, Zhao Zhangyuan, strongly opposed the construction of waste incinerators in China. The experts also advised Guangzhou Municipality to guarantee information transparency and provision of reliable information on the operation of waste incineration power plants. Moreover, experts suggested that local residents should be involved not only in the incineration power plant's EIA, but also in its daily operation. Following the outcomes and advice resulting from the forum, the Guangzhou municipal and Panyu district government kept quiet because of the Asian Games in Guangzhou, as argued by respondent 17. In November 2010, the Asian Games took place. In November 2010, the Ministry of Environmental Protection (MEP) of China added a new regulation to the incineration power plant construction standard, called Standard for Pollution Control on Municipal Solid Waste Incineration, implying that a decision on the location of a waste incineration power plant requires consideration of at least three alternative locations.[9]

Round 4: enhancing public participation for decision inputs (April–August 2011)

On April 19, 2011, the State Council (SC) clearly ruled that, for metropolitan areas with limited land resources and high population density, waste incineration power plants were the preferred option for processing waste.[10] On April 12, 2011, the Panyu district government, following information provided by the Guangzhou Urban Planning and Design Survey Research Institute, offered five alternative plant locations: Sansha (Dongyong Town), Basha (Lanhe Town), Xikengwei (Shawan Town), Xinlianer (Dagang Town), and Huijiang village (Dashi Street).[11] Furthermore, the Panyu district government ordered the Panyu Urban Management Bureau to carry out a public opinion poll before June 15. From April 13 to June 15, a public opinion poll was initiated by a local newspaper, *Southern Metropolitan Newspaper* (*Nanfang Dushi Newspaper*), through its online system, and local residents in Guangzhou were allowed to vote for the selection of the site for the Panyu waste incineration power plant. No detailed information was released about the response rates to the poll and who were involved. The questions asked could not be traced either. The result indicated that Dagang was the most preferred location. On August 15, the institute tasked with carrying out the EIA, the South China Institute of Environmental Sciences, formulated a ranking of locations for the waste incineration plant (see Table 5.1).[12] The intention for organizing a public opinion poll, however, was questioned by some citizens. Respondent 17 argued that Guangzhou Municipality indeed had already established Dagang as the alternative to Dashi, and the public opinion poll was not meaningful.

The South China Institute of Environmental Sciences' prioritization of locations, however, did not end the discussion on the plant's location, as three of the five proposed locations were quite close to Foshan City in the neighboring Shunde district. At roughly the same time, the Guangdong provincial government planned a new national economic zone in Guangzhou City, the so-called

Table 5.1 Public opinion poll and environmental assessment ranking for five potential locations

| Siting ID | Panyu | | Shunde | | Environmental assessment institute |
	Support	Object	Support	Object	Ranking
Dashi	45,374	57,966	1,300	12,411	3
Shawan	33,429	20,313	–	–	–
Dongyong	29,778	48,187	14,829	3,056	2
Lanhe	37,734	23,625	887	16,120	4
Dagang	53,740	27,469	607	18,540	1

Source: Data source for Panyu locals, http://gcontent.oeeee.com/1/ee/1ee1da76d31049b7/Blog/446/ee7809.html; data source for Shunde locals, www.shundecity.com/html/zt1/minyi2/index.html; data source for environmental assessment institute, www.scies.org/FileINFO.asp?Id=316.

Nansha New Area (*Nansha Xinqu*). Three of the five proposed locations, Lanhe, Dagang, and Dongyong, would be incorporated in this new economic area. Furthermore, the fourth alternative, Shawan, had been cancelled because of its short distance from a national nature reserve. Thus, local residents in Panyu district feared that Dashi would again be selected as the preferred location for the plant. However, with the assistance of Guangzhou Municipality, the Panyu district government reached an agreement with the Nansha district government (the main part of Nansha New Area) that the Panyu district government could freely choose the location for the incinerator, but it would process the waste from Nansha district as well. This agreement implies that the Panyu district government could further advance the incineration power plant construction in Panyu district, and that it could make its final choice from these alternatives.

Round 5: toward a decision (January 2012–April 2013)

On January 11, 2012, at the first meeting of the 14th Session of the People's Congress Conference of Guangzhou City, it was formally decided that a waste classification policy would be adopted to reduce the total amount of urban waste.[13] In order to reduce health hazards involved in the incineration of rubber, metal, glass, and leather, Guangzhou Municipality decided to further recycle and re-use through fostering the use of waste disposal facilities, fining citizens who refused to dispose of garbage according to the waste classification system, and charging citizens waste fees on the basis of weight. On April 6, at the eleventh meeting of the 14th Session of the People's Congress Conference of Guangzhou City, a set of new regulations on waste processing and re-usage was approved. Among them, both waste incineration and waste classification were determined as two crucial solutions for the urban waste problem.[14] Although Guangzhou Municipality established this policy, waste incineration in practice was still the only dominant approach. Respondent 19 claimed that waste classification was just a political slogan used by Guangzhou Municipality to deceive local residents by claiming that it had already put waste classification on its agenda. The truth, however, was that local governments in Guangzhou did not invest time and energy in implementing the waste classification policy. Even worse, they could intentionally mess it up. Respondent 19 additionally maintained that Guangzhou Municipality intentionally chose residential communities that were unlikely to complain as trial communities to implement its waste classification policy. In addition, respondent 19 argued that Guangzhou Municipality never explained to local citizens how the different types of waste should be classified.

In May 2012, the SC published a new notice on household waste disposal utility construction in 2015, clearly pointing out that the incineration approach would account for 35 percent of urban waste processing in China. In this proposal, it was also pointed out that incineration would account for 56 percent of the waste process in 2015 in Guangdong province.[15] In the same month, Guangzhou Municipality organized two symposiums to further consolidate the consensus on the urban waste process in Guangzhou. On May 18, the mayor of Guangzhou Municipality, Chen Jianhua, and 10 experts from Shanghai, Guangzhou, and other cities, publicly discussed the issue of urban waste management. Finally, this resulted in consensus on a policy. This policy implied "first classification, then collection, reduction, and ultimately harmless incineration, landfill, or biochemical processing."[16] Four days later, another symposium was held, in which 13 citizen representatives expressed their opinions to the mayor of Guangzhou Municipality on the disposal of urban waste. On June 19, the Guangzhou Urban Management Committee reported that, due to the mounting urban waste in Guangzhou City as well as public opinion on the urban waste disposal process, it had decided to build five rather than six waste incineration power plants within three years, all named *resource thermal power plants* (*Ziyuan Reli Dianchang*).[17] Moreover, Guangzhou Municipality established a sample plant, and it organized local citizens to have on-site visits to it. Many local Guangzhou citizens availed themselves of the opportunity to do so. Respondent 15 was one such visitor and argued that he believed that waste incineration was safe. Nevertheless, respondent 19 argued that the visits organized by Guangzhou Municipality were a persuading strategy and that waste incineration was not trustworthy.

Meanwhile, the importance of waste classification was highly emphasized by Guangzhou Municipality. It clearly pointed out that waste classification would be adopted in the whole of Guangzhou City before the end of 2012 and waste over-production charged with high waste process fees.[18] On July 10, 2012, Panyu district proposed Dagang as the first candidate for the construction of the waste incineration power plant, and prioritized Dongyong and Dashi in second and third place. Following this decision, the EIA was formally initiated, and two rounds of information disclosures and public hearings were organized. Detailed information about them is shown in Table 5.2.

On June 16, the Guangzhou Urban Management Committee recruited new members for a new commission, labeled Citizen Consultation and Supervision

Table 5.2 Environmental Impact Assessment (EIA) for the Panyu waste incineration power plant project

EIA	Main content
Information disclosure in planning EIA	The first information disclosure in planning EIA (May 20–31, 2011): it reported that the only waste landfill in Panyu district would stop operation in 2014. It thus was quite urgent to make a plan to construct a waste process facility to process the rising waste in Panyu district.
	The second information disclosure in planning EIA (August 15–26, 2011): the environmental assessment institute, South China Institute of Environmental Sciences, published the EIA report. It concluded that Shawan should be cancelled as a candidate, and Dagang, Dongyong, and Lanhe were chosen as the appropriate sites to construct waste incinerators.
Information disclosure in project EIA	The first information disclosure in project EIA (July 23–August 3, 2012): the Panyu waste incineration power plant was to be constructed in Dagang. The aim of the first information disclosure was to familiarize local residents with relevant information about the planned waste incineration power plant. It provided information about the reasons for location selection and technique selection.
	The second information disclosure in project EIA (November 19–29, 2012): it focused mainly on the assessment on the project's pollution disposal and environmental impacts on the surrounding environment. Public opinions were collected through a questionnaire, expert consultation, and a public forum.

Commission for Solid Waste Processing, which aimed to facilitate public participation in decision making on waste re-usage and waste reduction. On July 23, 2012, the first information disclosure meeting was held, and local residents were allowed to express their opinions through letter, fax, email, and telephone for 10 days. On August 4, this commission was officially established. Regarding the role of this commission in decision-making processes, different viewpoints were identified. Respondent 17, as one member of this commission, regarded the commission as a "flower vase" because it did not substantially influence the decisions of Guangzhou Municipality in decision-making processes about waste disposal. In addition, respondent 20 argued that some members of the commission were bought by Guangzhou Municipality.

The EIA for the Dagang waste incineration power plant approved by the Guangzhou Environmental Protection Bureau revealed publicly that the plant would be constructed by April 2013. In addition, Guangzhou Municipality claimed that the construction of future mega projects in Guangzhou must obtain the agreement of 75 percent of the residents from surrounding sensitive regions; but what constituted a sensitive region was still ambiguous. However, respondent 20 viewed this new regulation as a big advance in resolving Guangzhou's urban waste problem. Moreover, it was reported that local residents had already obtained compensation and that the whole village near the new proposed location, Dagang, would be moved to a new location. Because of this, few residents opposed the waste incineration power plant project. Different viewpoints were expressed by respondents about this decision to relocate the plant, however. Respondent 17 stated that the nearby residents were mostly villagers with low incomes, and the compensation from Guangzhou Municipality was attractive to them. Thus, they had no interest in opposing the construction of the waste incinerator. Respondent 19 argued that the new site had been used in the past for growing vegetables, and few villagers lived nearby. This implies that Guangzhou Municipality could construct a waste incineration power plant in Dagang at a very low cost. Some important dates and events are shown in Table 5.3.

The substantive, procedural, and institutional outcomes of the Panyu waste incineration power plant case

The Panyu waste incineration power plant was relocated from Dashi to Dagang. Guangzhou Municipality and the Panyu district government achieved their goal of constructing a waste incineration power plant, temporarily relieving the urgent waste problem in Panyu. Local residents in Dashi were winners too, as they had successfully moved the plant to another location. Local residents in Dagang obtained economic compensation from Guangzhou Municipality and would be relocated. They were winners as well. In short, this case might be viewed as a win–win game.

Regarding the outcomes at procedural level, this case lasted a long time, from 2009 to 2013. The conflict regarding the planning and construction of the Panyu waste incineration power plant became less adversarial because local governments in Guangzhou and local residents, including activists, had learned the importance of mutual collaboration in order to achieve a satisfactory resolution of the urban waste problem. The process seemed to be open over time (Johnson 2016). The other actors had more opportunities to be involved in formal decision-making processes. To a certain degree, the process in this case could be viewed as a good way of handing environmental conflicts.

As to the institutional outcomes, Guangzhou Municipality redesigned the institutions about waste disposal in Guangzhou. It began to attach high importance to waste reduction and waste re-use. In addition, a new commission was established by Guangzhou Municipality with the aim of engaging local residents in formal decision-making processes about waste disposal in Guangzhou. Finally, a new regulation about the construction of mega industrial projects was issued by Guangzhou Municipality, again in the hope of involving the public in

Table 5.3 Important dates and events in the Panyu waste incineration power plant case

Round No.	Important dates and events
Round 1 (September–October 2009)	• In September 2009, Guangzhou Municipality announced the construction of the Panyu incineration power plant in Dashi. • On October 30, 2009, the Panyu district government organized a press conference, in which four invited experts claimed that the incineration power plant would not result in pollution or health problems, and senior government officials from the Panyu district government explained the reasons for choosing Dashi as the preferred location for the plant.
Round 2 (November–December 2009)	• On November 23, 2009, residents assembled outside Guangzhou Municipality's building to protest against the incineration power plant construction. • On November 24, 2009, Guangzhou Municipality stated that construction would not take place if residents opposed it. • On December 10, 2009, the Panyu district government formally halted the Panyu waste incineration power plant.
Round 3 (February–November 2010)	• In February 2010, Guangzhou Municipality organized an expert forum. • In November 2010, the Asian Games were hosted in Guangzhou City.
Round 4 (April–August 2011)	• On April 12, 2011, the Panyu district government offered five alternative locations for the Panyu waste incineration power project. • On August 15, 2011, the EIA was published, and Dagang was chosen as the best location for waste incineration power plant construction.
Round 5 (April 2012–April 2013)	• On April 6, 2012, Guangzhou Municipality committed itself to waste classification and waste incineration. • On May 18, 2012, Guangzhou Municipality organized two symposiums. Government officials, experts, and representatives of local residents discussed the issue of the urban waste process. • On July 10, 2012, the Panyu district government selected Dagang as the location for incineration power plant construction. • In April 2013, the Dagang waste incinerator's EIA report was approved and the plant would start construction.

formal decision-making processes. These institutional changes reshaped the processes of waste disposal in Guangzhou, and it seems that local governments in Guangzhou have learned to resolve conflict – specifically referring to the construction of waste incineration power plants – collaboratively.

Strategies of local governments in Guangzhou during the Panyu waste incineration power plant case

In the first round, Guangzhou Municipality and the Panyu district government decided to construct a waste incineration power plant in Dashi. Local residents in Dashi were uninformed until Guangzhou Municipality publicly announced this. When the residents visited various government agencies in Guangzhou to express their disagreement to its construction, they received no response from these bodies. In addition, according to respondent 19, local governments in Guangzhou invited experts to persuade local residents of the harmlessness of waste incineration. These tactics were typical decide–announce–defend responses, characterized as a go-alone strategy. Then, when some unembedded activists organized collective activities, such as the Mask Show, they were summoned to the Panyu Police Department to pressurize them to maintain silence. This was coercion, characterized as a suppression strategy.

In the second round, local citizens in Dashi initiated a large-scale protest, and local governments in Guangzhou promised to halt the construction of the Panyu waste incineration power plant. This was project halting, characterized as a tension reduction strategy with the aim of avoiding any potential social disturbances.

In the third round, Guangzhou Municipality organized an expert forum in the hope of establishing a general guideline for waste disposal in Guangzhou. It is unclear how to categorize this response. It might be interpreted as a go-alone strategy because expert participation could be tactically used by Guangzhou Municipality to legitimize and bolster its initial decision, achieving waste incineration in Guangzhou. It could be interpreted as a facilitation strategy through which Guangzhou Municipality attempted to establish a platform to expedite the conflict resolution process. It could also be interpreted as a tension reduction strategy with the purpose of avoiding any possible social disturbances before the hosting of the Asian Games in Guangzhou. After the expert forum, local governments in Guangzhou kept silent and did not make any statements about the Panyu waste incineration power plant. Their silence can be categorized as a tension reduction strategy.

In the fourth round, about five months after the Asian Games in 2011, the Panyu district government came up with five locations for the Panyu waste incineration power plant project and organized a public opinion poll in order to legitimize the site selection. Although holding a public opinion poll should not be confused with authentic public participation, it could be stated that the scope for decision making was widened and new solutions were explored. This implies that local governments in Guangzhou provided more alternative options for local citizens in order to expedite the conflict resolution process. This was scope enlargement, characterized as a facilitation strategy.

In the fifth round, Guangzhou Municipality first established waste classification as an equally important policy as waste incineration for waste disposal in Guangzhou. In addition, it established a new supervision commission consisting of common citizens from a variety of social backgrounds. These responses could be viewed as changes in rules (or institutions) in order to enable the process of conflict resolution – specifically regarding the construction of waste incineration power plants in Guangzhou. The rule changes are characterized as institutional designs; this could be categorized as a facilitation strategy. Following this, local governments in Guangzhou organized symposiums in which local citizens were involved, and they compensated local residents in Dagang. These could be interpreted as a collaboration strategy; characterized as consensus seeking and compensation for a win–win solution, as argued by respondent 17, who stated that Guangzhou Municipality gradually learned to consult with local residents in decision-making processes about waste management. However, some skepticism was expressed by respondent 20, who said that the compromises made by local governments in Guangzhou were symbolic.

From the above analysis, an overview of strategies applied by local governments in Guangzhou is presented in Table 5.4.

The explanation for the application of government strategies in the Panyu case

Many choices of government strategies can be identified, as shown in Table 5.4. Nevertheless, four of them are chosen to be explained, as they reveal the general trend in the evolution of the Panyu case. They are: a go-alone strategy at the beginning, then suppression, tension reduction, and finally facilitation and collaboration at the end. In this section, the application of these four government strategies in the Panyu case is explained. In addition, the seven propositions drafted in Chapter 3 are used to explore the relationships between the individual conditions and the application of government strategies.

The explanation for the application of a go-alone strategy at the beginning

Initially, local governments in Guangzhou applied a go-alone strategy. They faced substantial pressure in resolving the waste problem in Guangzhou. They had established waste incineration as the main approach to disposal of waste, and, as early as 2006, they had already established Dashi as the location to construct a waste incineration power plant. They announced this in 2009, but some preparatory work had been done to advance its construction.

The explanation for the application of a go-alone strategy by local governments in Guangzhou was multi-faceted. First, waste incineration, established by the Chinese central government as a recycling industry, implies substantial economic benefits for industries. Respondent 19 argued that the national laws meant that the central government would grant the operators of waste incineration power plants substantial financial subsidies. If Guangri Corporation, affiliated to Guangzhou Municipality, could operate the waste incineration power plant in Panyu, it would obtain economic subsidies. This means that Guangzhou Municipality had a strong

Table 5.4 An overview of strategies applied by Guangzhou Municipality and the Panyu district government during the Panyu waste incineration power plant case

Actors	Round 1 (September–October 2009)		Round 2 (November–December 2009)		Round 3 (February–November 2010)		Round 4 (April–August 2011)		Round 5 (January 2012–April 2013)	
	Strategy	*Indicator*	*Strategy*	*Indicator*	*Strategy*	*Indicator*	*Strategy*	*Indicator*	*Strategy*	*Indicator*
Guangzhou Municipality and Panyu district government	Go-alone Suppression Go-alone	Decide and announce Coercion Defense and persuasion	Tension reduction	Project halting	Tension reduction go-alone/facilitation Tension reduction	Expert forum Silence	Facilitation	The widening of scope	Facilitation Collaboration	Changes in rules Relocation, consultation, and compensation

incentive to expedite waste incineration. Second, there was little opposition from others; local citizens were not well informed about the construction of this power plant. This also contributed to the application of a go-alone strategy.

The explanation for the application of a suppression strategy

Local governments in Guangzhou used state force to coerce the unembedded activists to keep them from organizing collective activities. This is characterized as a suppression strategy, which could be explained by three conditions: the involvement of unembedded activists, the occurrence of collective activities, and the use of social media.

1   *The involvement of unembedded activists*: After local governments in Guangzhou announced the construction of a waste incineration power plant in Dashi, some activists expressed their concerns about this decision. Respondent 20 claimed that most activists were unembedded. This implies that they had limited resources to mobilize to shape government decisions. They used some institutionalized approaches (such as letters and visits) as well as informal approaches (such as the Mask Show) to express their disagreement with the construction of a waste incineration power plant near their communities. For local governments in Guangzhou, the involvement of unembedded activists could not create sufficient pressure for them to make compromises. Consequently, the involvement of unembedded activists may have contributed to the application of a suppression strategy by local governments in Guangzhou.

2   *The occurrence of collective activities*: When unembedded activists organized collective activities (such as the Mask Show) to attract public attention, local governments in Guangzhou became nervous, because these activities could make them lose face or threaten their legitimacy. The application of a suppression strategy was a possible option for them to prevent this.

3   *The use of social media*: During this case, a blog message said that one senior government official in Guangzhou Municipality, who was strongly in favor of waste incineration, had a close relationship with the waste incineration industry, and it revealed that this official's brother and son were key figures in a waste incineration power plant company. Moreover, an activist, wearing a gas mask and holding a slogan "Opposing Waste Incineration and Protecting the Green Guangzhou City," walked around Guangzhou City center. Afterwards, she was taken away by the Panyu Police Office, and she wrote a post that was later widely reproduced by many popular online forums, ultimately causing the Panyu waste incineration power plant issue to become known around China. The activities organized by these unembedded activists could not be ignored by local governments in Guangzhou, as the information about waste incineration was widely known in Guangzhou and even around China. Following this, rumors could have triggered social disorder, and this was politically risky

for local governments in Guangzhou. They thus tended to apply a suppression strategy.

In sum, these three conditions are important in explaining the application of a suppression strategy by local governments in Guangzhou. Proposition 6 posited that local governments tend to apply a go-alone or a suppression strategy to deal with the involvement of unembedded activists, whereas they tend to adopt a tension reduction, a giving in, a collaboration, or a facilitation strategy to cope with the involvement of embedded activists. In this case, this proposition was confirmed: local governments may apply a suppression strategy to deal with the involvement of unembedded activists.

The explanation for the application of a tension reduction strategy

In this case, the Guangzhou Urban Management Commission's (GUMC) reception day opened a window of opportunity for local residents in Dashi to express their disagreement with constructing a waste incineration power plant near their community. On that day, hundreds of local Dashi citizens went to the GUMC. However, the GUMC did not provide a concrete solution to the conflict concerning the construction of the power plant. Respondent 16 argued that the GUMC did not have power to have a final say about the fate of the Panyu waste incineration power plant. He additionally claimed that Chinese governments do not have institutionalized policies to resolve complaints expressed by large-scale protesters. Regarding the waste incineration issue, many different government agencies were involved in deciding the fate of the plant. No single government agency has the absolute power to make decisions. And there is no agency to coordinate the resolution of such problems. As a result, local citizens were dissatisfied with the GUMC's performance, and they went to Guangzhou Municipality.

After that, Guangzhou Municipality applied a tension reduction strategy by temporarily halting the construction of the plant. The occurrence of a small-scale peaceful protest was an important condition in explaining the application of this strategy. The small-scale peaceful protest embarrassed Guangzhou Municipality, as argued by respondent 17. If it did not end the protest as soon as possible, its authority would be negatively threatened, following which local social order might be endangered. This situation was not tolerable for local governments in Guangzhou. As a result, they tended to apply a tension reduction strategy.

Proposition 1 in Chapter 3 was: local governments tend to apply a go-alone or a suppression strategy to deal with peaceful protests, whereas they tend to adopt a tension reduction or a giving in strategy to cope with violent protests. In this case, this proposition was disconfirmed and reformulated: local governments may apply a tension reduction strategy to deal with peaceful protests. Proposition 2 in Chapter 3 was: local governments tend to apply a tension reduction or a giving in strategy to cope with large-scale protests, whereas they tend to adopt a go-alone or a suppression strategy to deal with small-scale protests. This proposition was disconfirmed and reformulated: local governments may apply a tension reduction strategy to deal with small-scale protests.

After the occurrence of the small-scale peaceful protest, local governments in Guangzhou kept silent; but that did not last for long, and they organized an expert forum. This resulted mainly from the involvement of a national mass medium, CCTV. It reported the debates concerning the construction of the Panyu waste incineration power plant. Local governments in Guangzhou ended their silence. The CCTV report implied that the top leaders in the Chinese central government knew about the conflict concerning the construction of the Panyu waste incineration power plant. In addition, they might have been dissatisfied with the strategy applied by the Guangzhou local governments to handle this conflict. Afterwards, these local governments organized an expert forum. Unfortunately, it is uncertain to which strategy the expert forum belonged, but it is certain that the media criticism broke the silence of the local governments, as argued by respondent 17. As it is difficult to categorize the government response of the expert forum, proposition 3 cannot be confirmed, disconfirmed, or specified.

After the expert forum, a tension reduction strategy was applied by local governments in Guangzhou for a second time, primarily because of the occurrence of a planned event, the Asian Games (Lang and Xu 2013). The Asian Games were an important opportunity for the key leaders of local governments in Guangzhou. If they could host the games successfully, implying gaining face for the central government, they would be politically promoted. If they failed to do so, however, they would be punished through political demotion. Proposition 7 in Chapter 3 stated that local governments tend to apply a tension reduction strategy to cope with the occurrence of planned events, and it is uncertain which strategies are adopted by local governments to deal with unplanned events. In this case, this proposition was confirmed: local governments applied a tension reduction strategy in handling the occurrence of the planned event.

The explanation for the application of a collaboration and a facilitation strategy

At the end of the Panyu case, local governments in Guangzhou applied a collaboration and facilitation strategy at the same time. In general, three conditions were important in explaining this: the urgent nature of the waste problem in Guangzhou, the stage of the Panyu waste incineration power plant, and the involvement of unembedded activists.

First, *the serious urban waste problem in Guangzhou*: Guangzhou Municipality faced a serious waste problem. In 2011, household waste in Guangzhou City had reached 18,000 tons daily. This was estimated to rise to 22,700 tons every day in 2015. However, Guangzhou's waste disposal capacity had reached only 14,000 tons daily, thus exhibiting a great gap between waste production and disposal.[19] Local governments in Guangzhou therefore had a strong incentive to achieve the construction of waste incineration power plants in Guangzhou as soon as possible. During this case, local governments in Guangzhou increasingly recognized that few citizens would agree with the construction of waste incineration power plants near their communities, as argued by respondent 17. If local governments in Guangzhou continued to expedite the construction of waste incineration power plants as they did before in a top-down way, they would again attract strong opposition from local citizens, as stated by respondent 19. This might make it impossible to resolve the waste problem in Guangzhou. Therefore, it can be concluded that the urgent waste problem in Guangzhou contributed to the application of a collaboration and a facilitation strategy by local governments in Guangzhou.

Second, *the planning stage of the Panyu waste incineration power plant*: The Panyu waste incineration power plant was in its planning stage, implying that local governments in Guangzhou would not necessarily assume high costs if they relocated it. If they applied a go-alone or a suppression strategy to advance the construction of the waste incineration power plant in Dashi, local residents might organize another round of protests, which would endanger social order, implying high political costs for local governments in Guangzhou. The continuation of the waste incineration power plant project in Dashi was not a good option for them. Consequently, the fact that the Panyu waste incineration power plant was at the planning stage contributed to the application of a collaboration and a facilitation strategy by local governments in Guangzhou.

Third, *the involvement of unembedded activists*: Local governments in Guangzhou increasingly acknowledged the importance of unembedded activists during the process of resolving the waste problem. When Guangzhou Municipality identified one leading activist, Basuo Fengyun, it invited him to inspect a waste incineration power plant project in Macau to let him learn more about waste incineration; and it started to make contacts with some unembedded activists. Some of the unembedded activists in Guangzhou, such as Basuo Fengyun and A Jiaxi, were middle class and in non-state jobs. Different from activists at the bottom of Chinese society who are in a life-and-death struggle with governments, the activists in this case, although they were not embedded, owned their own homes and had no intention of fighting with local governments in a radically confrontational way. Rather, they cared about their quality of life, as respondent 19 claimed that "health is the most important issue and money is the second for me." She argued that, if Guangzhou Municipality agreed to construct a waste incineration power plant near its building, she would not oppose the construction of one near her community. In addition, these unembedded activists had substantial free time to seek "truth" based on which they could formulate evidence-based arguments to challenge or disqualify government decisions. In this case, after Guangzhou Municipality determined the temporary halting of the Panyu waste incineration power plant in Dashi, some embedded activists conducted on-site investigations into the influence on public health of a waste incineration power plant in Likeng. They organized blood tests for children living nearby. The results showed that these children had a higher level of lead content in their blood than average persons. Following this, a list of names was published anonymously, claiming that over 42 local residents living near the Likeng plant had died consequent to various cancers, mostly caused by waste incineration. This list received explosive media attention around China, and many residents in Guangzhou later became worried about the negative influence of waste incineration on their health. Subsequently, Guangzhou Municipality published a report to illustrate that the list was fraudulent. However, this report triggered more doubts among local citizens in Guangzhou about the negative influences of waste incineration on public health, as argued by respondent 17. These actions taken by unembedded activists greatly challenged the decision made by local governments in Guangzhou – the expediting of the Panyu waste incineration power plant. Therefore, these local governments tended to make compromises.

Moreover, the relationships between local governments in Guangzhou and unembedded activists changed during the case. An alignment relationship between them gradually formed. They interacted with one another frequently and then came to know one another. Some degree of interest alignment emerged during the case. As argued by respondent 20, "many people think we are enemies with government officials. But we are good friends, and we talk a lot about the waste problem." Guangzhou Municipality established a supervision commission, and some unembedded activists were recruited as members. Some activists, who initially opposed the government decision to construct the waste incineration power plant in Dashi, now had the opportunity to directly provide suggestions to Guangzhou Municipality about how the waste problem should be resolved. Respondent 17 claimed that he had drafted several proposals regarding the resolution of the waste problem in Guangzhou, and Guangzhou Municipality accepted some of his suggestions. Although local governments in Guangzhou did not institutionalize public participation in decision-making processes, some unembedded activists were granted more opportunities to provide inputs to government decisions. This conclusion, however, is arguable. Respondent 17 was satisfied with the actions taken by Guangzhou Municipality to facilitate public participation. However, respondents 19 and 20 argued that Guangzhou Municipality did a bad job in engaging public participation in resolving the waste problem. In short, the involvement of unembedded activists in formal decision-making processes implies that local governments in Guangzhou attempted to establish a collaborative relationship with them in order to jointly resolve the waste problem. Therefore, it could be concluded that the involvement of unembedded activists contributed to the application of a collaboration and a facilitation strategy at the end of the case.

The above explanation for the application of a collaboration and a facilitation strategy helped me to develop two propositions drafted in Chapter 3: proposition 4 and proposition 6. Proposition 4 was that local governments tend to apply a tension reduction, a giving in, a collaboration, or a facilitation strategy when projects are in their early stage, whereas they tend to adopt a go-alone or a suppression strategy when projects are in their late stage. In this case, this proposition was confirmed: local governments may apply a collaboration and a facilitation strategy to deal with debated projects in their early stage. Proposition 6 in Chapter 3 was that local governments tend to apply a go-alone or a suppression strategy to deal with the involvement of unembedded activists, whereas they tend to adopt a tension reduction, a giving in, a collaboration, or a facilitation strategy to cope with the involvement of embedded activists. This proposition was disconfirmed and reformulated: local governments may apply a collaboration or a facilitation strategy to deal with the involvement of unembedded activists.

Discussion

In the Panyu case, both a collaboration and a facilitation strategy were applied by local governments in Guangzhou. It is assumed that the findings in this case show some generic characteristics of the well-developed regions of China regarding the governance of environmental conflicts. This case is useful for providing insight into the application of government strategies – especially the emergence of the collaboration and the facilitation strategy in a Chinese context. It is interesting to identify some generic conditions in this case that are helpful in explaining the emergence of these strategies (see Li et al. 2016a).

First, this case occurred in Guangzhou, located in one of the most prosperous regions in China. A large group of middle-class citizens (such as lawyers, journalists, and intellectuals) live there and have a strong rights consciousness, as argued by respondent 20. They tend to use evidence-based information (such as scientific research by independent third parties or national laws) to argue with local governments in order to show their disagreement with, and concerns about, government decisions. This creates an environment for negotiation between local governments and other actors, implying that local governments in Guangzhou have a relatively higher possibility of collaborating with local citizens (Bie, Jong, and Derudder 2015).

Second, Guangzhou is rather close to Hong Kong, and many governance practices regarding problem solving are transplanted by local Guangzhou governments. Governments thus tend to have relatively open attitudes to external critics, as argued by respondents 17 and 20. This again fosters a climate for deliberation and negotiation between various disputants involved in this case; this facilitated the emergence of a collaboration and a facilitation strategy.

Third, Guangzhou is relatively far away from Beijing, the political center of China. Local mass media in Guangzhou, for instance, have a relatively greater degree of freedom. Consequently, Guangzhou has a comparatively open political climate, and this is helpful for the development of civil society there. As argued by respondent 20, local governments in Guangzhou are relatively open to public criticism. Many local citizens in Guangzhou can publicly criticize government agencies, a tradition special to Guangzhou. The open political climate in Guangzhou also contributed to the emergence of a collaboration and a facilitation strategy applied by local governments in Guangzhou.

Conclusions

The Panyu waste incineration power plant case in Guangzhou was analyzed in this chapter. The theoretical framework established in Chapter 3 was used to address two issues in this case study: the application of government strategies and their explanation.

Regarding the application of government strategies, four government strategies were adopted by local governments in Guangzhou. Initially, they applied a go-alone strategy. They advanced the construction of the Panyu waste incineration power plant without informing local residents nearby. When activists and local residents organized collective activities to show their concerns about the construction of the plant, local governments applied a suppression strategy. After this, a small-scale peaceful protest occurred, and local governments in Guangzhou claimed that the plant would not be further advanced; this is characterized as a tension reduction strategy. Finally, they applied a combination of a collaboration and a facilitation strategy.

The application of the four government strategies during the Panyu case was explained using seven conditions: the form of protest, the scale of protest, the stage of the project, the position of the national mass media, the position of higher-level governments, the involvement of activists, and the occurrence of events. In general, three conditions are really important in explaining the application of government strategies in this case: the early stage of the Panyu waste incineration power plant, the involvement of unembedded activists, and the absence of support from higher-level governments and national mass media. In addition, three propositions drafted in Chapter 3 were disconfirmed and reformulated.

1   Proposition 1: local governments may adopt a tension reduction strategy to deal with peaceful protests.
2   Proposition 2: local governments may apply a tension reduction strategy to cope with small-scale protests.
3   Proposition 6: local governments may apply a collaboration and a facilitation strategy to deal with the involvement of unembedded activists.

Finally, this case study has shown that some other conditions, such as the use of social media and the urgency of the (waste) problem, might also influence the application of government strategies in environmental conflicts.

Notes

1   http://gcontent.oeeee.com/a/7d/a7d8ae4569120b5b/Blog/d3d/c53c90.html, available on August 14, 2012.
2   www.infzm.com/content/39868, available on August 14, 2012.
3   See Appendix 2 for the respondents' profiles and place of interview.
4   http://news.sina.com.cn/c/sd/2011-10-25/111223358836.shtml, available on August 3, 2015.

5 http://news.sina.com.cn/c/sd/2011-10-25/111223358836.shtml, available on August 3, 2015.

6 http://news.sina.com.cn/c/sd/2011-10-25/111223358836.shtml, available on August 3, 2015.

7 http://news.xinhuanet.com/local/2009-10/31/content_12364707.htm, available on October 8, 2013.

8 www.chinanews.com/gn/news/2009/11-26/1984592.shtml, available on October 23, 2015.

9 www.mep.gov.cn/gkml/hbb/bgth/201011/t20101125_197974.htm, available on October 8, 2013.

10 http://finance.sina.com.cn/roll/20120606/021612231982.shtml, available on October 9, 2013.

11 http://news.nfmedia.com/nfdsb/content/2011-10/17/content_31527251.htm, available on October 8, 2013.

12 www.scies.org/FileINFO.asp?Id=316, available on October 8, 2013.

13 www.gz.gov.cn/publicfiles/business/htmlfiles/rdzxlhzt/gdtp/201201/892431.html, available on October 8, 2013.

14 http://news.ycwb.com/2012-04/07/content_3763756.htm, available on October 8, 2013.

15 www.gov.cn/zwgk/2012-05/04/content_2129302.htm, available on October 9, 2013.

16 http://gd.people.com.cn/n/2012/0518/c123932-17051708.html, available on October 8, 2013.

17 http://news.xinhuanet.com/energy/2012-06/19/c_123303207.htm, available on August 14, 2012.

18 http://informationtimes.dayoo.com/html/2012-04/07/content_1664984.htm, available on August 14, 2012.

19 www.infzm.com/content/74360, available on August 7, 2015.

6   Government strategies in governing environmental conflicts

The Dalian PX plant as a single case

Introduction

Dalian is a port city in northeast China, with over six million residents living there in 2009. It has widely been recognized as one of the cleanest and most livable cities in China. Meanwhile, Dalian is a traditional petrochemical base in China, and many mega industrial projects have been constructed and operate there. Dalian Municipality had a strong economic reliance on these industries. In 2011, this alignment relationship between Dalian Municipality and local industries were threatened due to the occurrence of a large-scale protest involving about 12,000 participants. This protest shocked both the Chinese government and citizens, as a protest on such a scale has been very rare in recent years. The participants involved were dissatisfied with the operation of a PX plant in Dalian given its potential negative harm on their health. Dalian Municipality immediately faced substantial pressure from the explosive attention all around China. One intriguing issue that warrants further study is how Dalian Municipality coped with this conflict. This chapter reports an in-depth case study with the aim of identifying which strategies were applied by Dalian Municipality in dealing with the conflict regarding the operation of the PX plant and explaining why these strategies were applied. This chapter proceeds in five sections. First the Dalian PX case is introduced and then the strategies adopted by Dalian Municipality are set out. Four key government strategies are identified, one of which – tension reduction – is used at two different stages of the project: a go-alone, a suppression, a tension reduction, a giving in, and again a tension reduction strategy. The next section explains why Dalian Municipality applied these five different strategies. The discussion appears in the following section and, finally, conclusions are drawn.

The Dalian PX case

The Dalian PX case is introduced in detail in this section, broken down into four subsections: background, network character, process, and outcome of the case.

Background to the Dalian PX case

Paraxylene (PX), a chemical substance, is used primarily as a basic raw material in the manufacture of terephthalic acid (TPA), purified terephthalic acid (PTA), and dimethyl-terephthalate (DMT). TPA, PTA, and DMT can be used to manufacture polyethylene terephthalate (PET) saturated polyester polymers. Polyesters are used to produce fibers and films. China is the biggest PX production and consumption country in the world. From 1992 to 2005, the annual production of PX increased by 11.59 percent annually, and its consumption in China grew by 16.36 percent. The plan to construct a PX plant in Dalian was initiated in 2003, with a state-owned enterprise (SOE), Dalian Petrochemical Company, technically responsible for it. However, this SOE was engaged in costly site relocation and could not amass sufficient money for investment in a PX plant at that time. In 2004, the State Council (SC) issued the Decision on the Reform of the Investment System, providing private enterprises the same investing opportunities as SOEs. This gave the Dalian State-Owned Assets Supervision and Administration Commission (SASAC) the legitimacy to involve Fujia as the main shareholder in the PX project in 2005.

On September 28, 2005, the Fujia PX project was formally approved by the State Environmental Protection Administration (SEPA). In December, 2005, the National Development and Reform Commission (NDRC) approved the construction of the Fujia PX project. In addition, the Fujia PX project was established by the NDRC as a key project in the Eleventh Five-Year Plan (2006–2010).[1] The Fujia PX plant is about 20 kilometers away from the city center, and few people noticed its construction and operation. No public hearing was reported to have been held in relation to the acceptability of the Fujia PX project in 2005. Local mass media in Dalian did not report the possible negative effects of the Fujia PX project either.

The network context of the Dalian PX case

Regarding the network character of the Dalian PX case, some general institutional information is first presented. As early as 2003, the National Revitalization Strategy of the Old Industrial Bases in Northeast China, established by the SC, already listed the petrochemical industry as a priority for the region's economic development. Dalian is part of that region. This implies that Dalian Municipality had the priority of developing its local economy through industrial development. In addition, the gap between the production and the consumption of PX in China had reached 1.55 million tons in 2005. The short supplies of PX pushed the NDRC to decide to promote this industry in China. The prospect of substantial economic profits from the PX industry made Dalian Municipality eager to have a PX plant in its jurisdiction.

The Dalian SASAC, a bureau of Dalian Municipality, is responsible for managing SOEs in the city. In September 2005, it signed a contract with Fujia Company, a private enterprise, to establish a new joint company, Fujia Dahua

Petrochemical Co., with its subordinate SOE, Dalian Petrochemical Company, as the minority shareholder. The Fujia Dahua Petrochemical Company owns the Fujia PX project, which produces 0.7 million tons of PX and contributes 0.9 billion RMB in tax annually. The chief executive was once a real estate businessman. Because the Fujia PX plant was to have an annual PX production over 100,000 tons, it had to obtain planning and operation approval from the NDRC, a high-level governmental commission under the SC, hierarchically at the same level as ministries, but in practice more highly regarded. Before the formal construction of the Fujia PX plant, it also had to be approved by SEPA before 2008. On April 5, 2007, the SC promulgated the first decree about information disclosure. Following this, SEPA enacted its environmental information disclosure measure.

Several different actors can be identified in the Dalian PX case: Dalian Municipality, NDRC, SEPA, the environmental protection agency of the Liaoning provincial government, port management agencies, and local citizens. Three key actors can be identified, namely, Dalian Municipality, local residents in Dalian, and the Fujia PX plant. The Fujia PX plant, as the project operator, had a strong incentive to advocate the construction and operation of the PX plant with the purpose of advancing its economic interests. Dalian Municipality was highly supportive of the Fujia PX plant because of its substantial economic benefits to local development. Local residents in Dalian wanted to stop the operation of the Fujia PX plant and remove it from Dalian given its potentially negative influence on their health.

The process of the Dalian PX case

The Dalian PX case is differentiated into four rounds. Criteria to distinguish rounds have to do with important decisions or events that result in a shift in issues being discussed, or that affect the nature of interactions. Five crucial junctures that demarcate the four rounds are identified. They are presented in the following.

1   In June 2009, the Fujia PX plant started formal operation without fully informing local citizens in Dalian. This was the first crucial decision as it showed the closed decision-making processes of Dalian Municipality and signified the beginning of the first round (or the beginning of the case).
2   In 2009, an embedded activist, Lu Renzi, was suppressed when he expressed his concerns about the operation of the PX plant. This crucial decision showed the characteristic of interaction (the application of a suppression strategy by Dalian Municipality) between Dalian Municipality and the other actors. This crucial decision signified the end of the first round.
3   On August 8, 2011, the plant's protective dyke was damaged by Typhoon Muifa, prompting the citizens in Dalian to become worried, and they planned to initiate a large-scale protest. This event triggered the mobilization of citizens to oppose the operation of the PX plant. It signified the end of the second round.

4   On August 14, 2011, the mayor of Dalian Municipality claimed that the PX plant would be removed consequent to the large-scale protest initiated by the local citizens. This was a crucial decision, showing the change in inter-actions between Dalian Municipality and local citizens (the application of a giving in strategy by Dalian Municipality). It signified the beginning of the third round.

5   In December 2012, the Fujia PX plant was still in operation. This was a crucial decision that showed the substantial outcome of the case. It signified the end of the whole case.

In what follows, the four rounds of the case are described in detail.

Round 1: advancing the operation of the PX plant (before July 2010)

In October 2007, the construction of the Fujia PX plant formally began in the Petrochemical Industrial Park. In June 2009, the plant started formal operation. However, this PX project did not get approval from the Liaoning Provincial Environmental Protection Department until April 2010. This implies that the Fujia PX project began to operate before receiving formal approval.[2] Right after the plant went into full production, Dalian Municipality's official newspaper, *Dalian Daily*, publicly announced that, from being an oil refining base, Dalian was now witnessing a shift toward becoming a booming petrochemical base.[3]

In 2009, Lu Renzi, a reporter from Dalian Television, published an open letter to top officials on Dalian online. On behalf of Dalian citizens, he demanded that Dalian Municipality organize a press conference to answer citizens' questions about the operation of the PX plant in Dalian. However, without an official response, Lu Renzhi, who had a high profile in social media, ended up by resigning from his job, reportedly under pressure.[4] On July 16, 2010, there was an explosion in an oil storage depot belonging to China National Petroleum Corporation (CNPC), as a result of which 1,500 tons of oil spilled into the Yellow Sea.[5] There were quite a number of oil storage depots near the accident scene, and any further spread of the incident would have been a disaster for the residents living nearby. Terrified citizens fled from their homes, blocking the highways out of Dalian. Eventually, the fire was extin-guished, but one firefighter died.

Round 2: reducing the negative effect of the dyke damage at the PX plant (July–August 2011)

On July 16, 2011, a fire occurred on the CNPC site. Although it occurred in the CNPC plant and not in the Fujia PX plant, the citizens in Dalian speculated in panic on the dreadful impacts if the Fujia PX project were to explode too. On August 8, 2011, Typhoon Muifa struck Dalian and breached one of the PX plant's protective dykes. Rumors spread on social media that a leakage from the PX storage tanks would flood the Yellow Sea with highly toxic PX. Accompanied by several senior government officials from Dalian Municipality, a China Central Television (CCTV) news program intended to investigate this accident but failed to access the site because of resistance from the plant's executives.[6] Even worse, some staff from the Fujia PX plant attempted to snatch the reporters' cameras. As a result, violent conflicts occurred between the Fujia PX plant staff and the reporters from CCTV. When CCTV still intended to broadcast the story, the program director received a request to pull the segment and did so.[7] This news spread widely online, and many popular online forums in China posted a message calling local residents in Dalian to join a large-scale demonstration on August 14. Meanwhile, Dalian Municipality promised to take more administrative measures to avoid further accidents.[8] The Dalian Port Company, an SOE, echoed this request by immediately investing 100 million RMB in its emergency facilities. On August 9, 2011, Dalian Municipality publicly stated that related information regarding the removal of the Fujia PX project would be released soon.

Round 3: stopping the operation of the Dalian PX plant (August–September 2011)

On the morning of August 14, 2011, about 12,000 people assembled in the city center to require Dalian Municipality to stop the PX project and provide an accurate timeline for its removal from Dalian. When local residents walked around the city center, many police from the Dalian Police Department attempted to stop this. No violent actions occurred. The new mayor of Dalian, Tang Jun, had a direct conversation with the protest participants using a loudspeaker, and he promised to stop the Fujia PX project immediately. That night, Dalian Municipality organized an emergency meeting to discuss the removal of the Fujia PX plant.[9] The next day, August 15, many residents in Dalian received a letter from Dalian Municipality to inform them of the harmlessness of the PX plant. On August 16, Dalian Municipality promised to draw up a proposal to remove the Fujia PX project. On September 9, it stated that the Fujia PX project had stopped production. On September 26, 2011, reflecting on the Dalian PX protest, the Industry Coordination Department of the NDRC co-signed with four ministries (the Ministry of Industry and Information Technology of China, the Ministry of Environmental Protection of China, the State Administration of Work Safety of China, the Ministry of Land and Resources of China) an emergency notice titled "Strengthen Security of Sensitive Products like PX." In this notice, the protest that had occurred in Dalian was deemed by the government to be due to the unsafe production of PX.[10]

Round 4: resumption of the Fujia PX plant (October 2011–December 2012)

After the large-scale protest in Dalian, PX became a forbidden keyword on social media, and no news reports have been available on the Fujia PX plant's on-going circumstances since then. In October 2011, one local mass medium in

Dalian, *Dalian Diary*, published some comments by some experts from Beijing that concluded that the Fujia PX project was not as dangerous as people thought.[11] In November 2011, Dalian Municipality publicly claimed that the Fujia PX plant would be removed in an orderly way. On December 8, 2011, it organized a meeting to discuss the plan to remove the PX plant. In this meeting, the mayor of Dalian Municipality publicly stated that all relevant petrochemical enterprises would be removed to Changxing Island Industrial Park.[12] However, a commercial newspaper in Beijing became aware of, and reported, an official reply letter to the Dalian customs, implying Dalian Municipality's internal approval for the Fujia PX project to resume production.[13] In August 2012, a regulation Interim Measures on Social Stability Risk Assessment for Major Fixed Asset Investment Projects released by NDRC determined that a social stability risk assessment would be a precondition for NDRC to approve large projects. Reflecting on the PX protests, the NDRC intended to establish an institutional channel to enhance public participation before the approval of large industrial plants in order to avoid potential social unrest. In December 2012, it was reported that the Fujia PX project was still in operation,[14] and some even said that the Dalian PX project had never stopped operation, because of the high costs of removing the PX plant to another location. It has been reported that the annual PX sales for the Dalian PX plant are worth about one billion dollars, and that stopping production for one month would imply a loss of about 80 million dollars.

In conclusion, the Dalian PX case evolved in four rounds. Important dates and events are presented in Table 6.1.

The substantive, procedural, and institutional outcomes in the Dalian PX case

At the end of the Dalian PX case, Dalian Municipality determined to relocate the Fujia PX plant, although this failed to be implemented. Dalian Municipality's statement about relocating the plant could be interpreted as a tension reduction strategy, as indicated by the delaying of the project relocation with a purpose of diverting public attention from it. To date, it is not easy to judge who the winner was and who the loser was. Although the resumption of the Fujia PX plant might yield substantial tax incomes for Dalian Municipality, the municipality suffered severe losses of legitimacy and authority. Local residents in Dalian seemed to be the losers, but their actions during the case might teach Dalian Municipality to be prudent about its application of strategies in the future.[15] The Fujia PX plant did not suffer significant economic losses in this case. However, its social reputation was damaged. Overall, there may be no winners in this case.

Regarding the outcomes at procedural level, the decision-making processes in the case were mostly closed and other actors – mostly local citizens in Dalian – had few opportunities to influence the decisions made by Dalian Municipality in coping with the conflict regarding the operation of the PX plant. In addition, the costs for conflict resolution were very costly for Dalian Municipality as the

Table 6.1 Important dates and events in the Dalian PX case

Round No.	Important dates and events
Round 1 (before July 2010)	• In June 2009, the Fujia PX plant started formal operation. • In 2009, Lu Renzi, a reporter from Dalian TV, published an open letter to top officials on Dalian online. On behalf of Dalian citizens, he suggested halting the Fujia PX project, but later he quit his job. • On July 16, 2010, there was an explosion at an oil storage depot belonging to China National Petroleum Corporation, as a result of which 1,500 tons of oil spilled into the Yellow Sea.
Round 2 (July–August 2011)	• On July 16, 2011, a fire occurred on the CNPC site. • On August 8, 2011, Typhoon Muifa breached a dyke near the Fujia PX plant. • On August 9, 2011, Dalian Municipality publicly stated that information regarding the removal of the Fujia PX project would be released soon.
Round 3 (August–September 2011)	• On August 14, 2011, 12,000 people went to Dalian city center to demand that Dalian Municipality stop the PX project. • On the afternoon of August 14, 2011, Dalian Municipality promised to remove the PX project. • On August 16, 2011, Dalian Municipality promised to make a proposal to remove the Fujia PX project.
Round 4 (October 2011–December 2012)	• In November 2011, Dalian Municipality publicly claimed that the Fujia PX plant would be removed in an orderly way. • In December 2012, it was reported that the Fujia PX project was still in operation.

halting or the removal of the PX plant implies substantially high costs for it. In short, the process of the Dalian case was far from best practice in governing environmental conflicts, as collaboration and negotiation did not emerge in this case.

As for the outcome at institutional level, the Chinese central government established social risk assessment as a prerequisite procedure for the formal approval of any mega industrial plants in China. However, its implementation and institutionalization are still unclear.

Government strategies in the Dalian PX case

In the first round, the Fujia PX plant started operation before receiving formal approval from the Environmental Protection Department of the Liaoning provincial government. After Dalian Municipality announced that the plant had started production, it mobilized local mass media to promulgate the plant as an economically beneficial project for Dalian. This shows Dalian Municipality's strong intention to realize the construction and operation of the plant. This was a traditional *announce–defend* style of response, characterized as a go-alone strategy. Then, when an embedded activist from Dalian Television publicly questioned the legitimacy of the plant, he was later obliged to quit his job. This shows that Dalian Municipality attempted to suppress different opinions and actions; this is characterized as a suppression strategy.

In the second round, after the dyke damage in the Fujia PX plant, Dalian Municipality promised to take some administrative measures to avoid the occurrence of explosions in future in the hope of relieving the worries of local residents. Meanwhile, it claimed that it would investigate the Fujia PX plant and that detailed information about the removal of the plant would be released. These responses imply that Dalian Municipality attempted to calm local residents down in order to avoid the occurrence of large-scale protests initiated by local citizens; this was a tension reduction strategy.

In the third round, the mayor of Dalian Municipality first had a face-to-face dialogue with the participants in the large-scale protest. Then he promised to stop the operation of the Fujia PX plant and remove it to another location. The halting and relocation of the Fujia PX plant is characterized as a giving in strategy. Afterwards, Dalian Municipality organized an emergency meeting to show the public how it was trying to implement its promise to move the plant, and it delivered letters to the public to stress the harmlessness of waste incineration. These responses taken by Dalian Municipality were aimed at calming local residents down, as argued by respondent 24; this is characterized as a tension reduction strategy.

In the fourth round, keywords relating to the Fujia PX plant were determined by Dalian Municipality to be sensitive words on social media, and relevant information about the Fujia PX plant was blocked. This information blockage is characterized as a suppression strategy. Then, Dalian Municipality publicly claimed that the Fujia PX plant would be removed to another location. At the same time, the PX plant resumed operations secretly. These responses are characterized as a tension reduction strategy.

The strategies adopted by Dalian Municipality during the Dalian PX case are summarized in Table 6.2.

The explanation for the application of government strategies in the Dalian PX case

Table 6.2 shows that, over time, Dalian Municipality adopted several different strategies in handling the conflict concerning the operation of the Fujia PX plant. Five key government strategies were applied by it, showing the general evolution of the case. They are the application of a go-alone, a suppression, a tension reduction, a giving in, and again a tension reduction strategy. These five choices are explained in the following five subsections.

Table 6.2 An overview of strategies applied by Dalian Municipality during the Dalian PX case

Actor	Round 1 (before July 2010)		Round 2 (July–August 2011)		Round 3 (August–September 2011)		Round 4 (October 2011–December 2012)	
	Strategy	Indicator	Strategy	Indicator	Strategy	Indicator	Strategy	Indicator
Dalian Municipality	Go-alone	Announcement, and media framing	Tension reduction	Reactive remedy and official promise	Giving in	Relocation	Suppression	Information blocking
	Suppression	Firing the activist			Tension reduction	Emergency meeting and persuasion	Tension reduction	Project resumption secretly

The explanation for the application of a go-alone strategy

Initially, Dalian Municipality announced the construction of the Fujia PX plant, about which local citizens had not been fully informed. This is characterized as a go-alone strategy. Two conditions are important in explaining this strategy: the support of higher-level governments and the absence of strong opposition from other actors.

The support of higher-level governments for the PX plant was the first important condition that contributed to the application of the go-alone strategy. In this case, the Fujia PX plant was highly favored both by the Chinese central government (the NDRC) and the Liaoning provincial government (mainly the Environmental Protection Department of the Liaoning provincial government). The NDRC designated the Fujia PX as a key industrial project to promote the national economy. The Environmental Protection Department of the Liaoning provincial government tolerated the operation of the Fujia PX plant before it received approval for its environmental impact assessment (EIA). The support of higher-level governments gave Dalian Municipality confidence to further advance the PX plant. Therefore, it can be concluded that the support of higher-level governments contributed to the application of a go-alone strategy by Dalian Municipality.

In addition, the absence of opposition of the go-alone strategy in this case. At the beginning of the case, local residents were not fully informed that the PX plant would be constructed in Dalian. As a result, they did not voice concerns about the construction of the plant, and consequently there was little pressure on Dalian Municipality to take their worries seriously.

Because of the high support from higher-level governments and the small opposition from other actors, Dalian Municipality tended to apply a go-alone strategy. This conclusion helped to confirm proposition 5 in Chapter 3. Proposition 5 was that local governments tend to adopt a tension reduction, a giving in, a collaboration, or a facilitation strategy when higher-level governments criticize debated projects, whereas they tend to apply a go-alone or a suppression strategy when higher-level governments support debated projects. In this case, proposition 5 was confirmed: local governments may apply a go-alone strategy when higher-level governments support the continuation of debated projects.

The explanation for the application of a suppression strategy

In the second round of this case, Dalian Municipality applied a suppression strategy to fire an embedded activist when he expressed his concerns about the operation of the PX plant.

Although Lu was embedded, he had limited administrative power or political influence to challenge the decision made by Dalian Municipality – the construction of the Fujia PX plant. However, the application of a go-alone strategy was not attractive for Dalian Municipality, because he was a popular figure in social media. If he continued to publicly express his opposition to the Fujia PX plant, local citizens might first become aware of his viewpoint, then accept it, and finally collectively take to the streets to oppose the plant's operation. Dalian Municipality thus had to take action immediately in order to minimize his influence on public opinion. A suppression strategy was one option. No detailed information and data are available to ascertain whether Dalian Municipality took other actions (such as persuasion or coercion) to silence Lu. What can be seen is that Lu quit his job; this was a relatively serious punishment for an embedded person in China, as argued by respondent 8. Normally, if embedded persons do something that might endanger political authority or social order, governments often order their managers to warn or persuade them to keep silent.

It can be concluded that the involvement of an embedded activist contributed to Dalian Municipality's application of a suppression strategy. Proposition 6 in Chapter 3 was: local governments tend to apply a tension reduction, a giving in, a collaboration, or a facilitation strategy to cope with the involvement of embedded activists, whereas they tend to apply a go-alone or a suppression strategy to deal with the involvement of unembedded activists. In this case, this proposition was disconfirmed and reformulated: local governments may apply a suppression strategy to deal with the involvement of embedded activists.

The explanation for the adoption of a tension reduction strategy

During the Dalian PX case, three events occurred (an explosion, a fire, and dyke damage). After these, Dalian Municipality claimed that it would take action to avoid the occurrence of incidents at the PX plant; this is characterized as a tension reduction strategy.

In this stage, a go-alone, a suppression, or a giving in strategy was ineffective for Dalian Municipality to cope with the occurrence of events – a fire and dyke damage. A go-alone strategy was not attractive for Dalian Municipality as rumors about the potentially negative effects of the PX plant on public health had already been widely disseminated. If the rumors continued to be disseminated around Dalian, worried citizens might take contentious actions to oppose the operation of the plant, after which social order would be threatened. A suppressions strategy (such as coercion, state suppression, or information blockage) might result in more serious panic among local citizens, who would think Dalian Municipality was attempting to intentionally cover up the truth about the dreadful impact of the Fujia PX plant. This would trigger more distrust of Dalian Municipality among citizens, implying a high potential of social unrest. A giving in strategy was not a feasible option for Dalian Municipality either. Giving up the Fujia PX plant implied not only that Dalian Municipality would suffer great economic losses, but also that it might have to assume the extremely high costs of removing the Fujia PX plant. Finally, Dalian Municipality adopted a tension reduction strategy.

In general, two conditions were crucially important in shaping the application of the tension reduction strategy: the occurrence of unplanned events and the involvement of national mass media.

First, the occurrence of unplanned events – explosions at another location and dyke damage of the PX plant – attracted substantial public attention on the operation of the Fujia PX plant, and the local residents in Dalian started to worry about its negative effect on their health. The explosive public attention triggered the spread of rumors about the negative effects of PX plants on public health. This might have threatened social order. The application of a tension reduction strategy was useful to relieve the worries of local residents in Dalian and avoid the occurrence of social instability.

Second, a national mass medium, CCTV, investigated the Fujia PX plant. It did not publicly show its position on the strategies applied by Dalian Municipality. However, many citizens around China knew that the Fujia PX plant had been investigated by CCTV, and this created high pressure for Dalian Municipality, which had to become prudent about the continued operation of the plant. Consequently, a tension reduction strategy seems an appropriate option for Dalian Municipality in this case.

Proposition 7 in Chapter 3 posited that local governments tend to apply a tension reduction strategy to deal with the occurrence of planned events, and that it is difficult to judge the directions of government strategies in dealing with unplanned events. In this case, this proposition was specified and reformulated: local governments may apply a tension reduction strategy to cope with the occurrence of unplanned events.

The explanation for the application of a giving in strategy

After the occurrence of the large-scale peaceful protest in Dalian, the mayor of Dalian Municipality publicly claimed that the Fujia PX plant would be removed. This is characterized as a giving in strategy.

The application of a go-alone, a suppression, a tension reduction, a collaboration, or a facilitation strategy at this stage was inappropriate for Dalian Municipality as the protest was large scale. The use of a go-alone strategy was inappropriate since it might result in social instability. State repression was not an option for Dalian Municipality in this stage either.[16] First, repressing a protest involving over 12,000 participants was politically risky for Dalian Municipality. If it used state repression to handle this protest, the conflict might spiral out of control and social disorder might follow. Dalian Municipality would have been punished for this by the central government. Second, the nonviolent nature of the protest implied that the involved participants did not intend to disrupt social order or challenge state authority; this was confirmed by respondents 21 and 22. It was unlikely that Dalian Municipality would interpret the protest as anti-state oriented, and this further disqualified the use of state force. Third, the use of state suppression might be exposed by the participants through social media, and this would result in social condemnation. It was highly possible that local citizens would again resort to using social media to expose the actions taken by Dalian Municipality, and this in turn made the municipality prudent about the use of state force. Fourth, a national mass medium, CCTV, had conducted an investigation into the Fujia PX plant. This meant that the top leaders of the central government had noticed the conflict over the operation of the plant. It was politically risky for Dalian Municipality to use state force to suppress citizens.

A tension reduction strategy was not appropriate either. Before the occurrence of the large-scale protest, the municipality had publicly stated that it would take measures to avoid the occurrence of incidents at the plant in the hope of avoiding protest. However, a tension reduction strategy was not appropriate, as local citizens did not trust the statement made by Dalian Municipality, and a large-scale protest occurred as planned. Given citizens' distrust of Dalian Municipality, the application of a tension reduction strategy for a second time would have been ineffective in ending the protest immediately. Finally, facing such a large-scale protest initiated by local citizens, a collaboration or a facilitation strategy would also have been ineffective. After all, joint fact-seeking based on collaboration and facilitation to achieve a solution acceptable to all actors would have required a lot of time.

A giving in strategy was thus an appropriate option for Dalian Municipality in this stage to address the large-scale peaceful protest. Respondents 21 and 22 claimed that there were no organizers and that they were activated by anonymous announcements from social media. This information was also confirmed by respondents 24 and 25. The mayor of Dalian Municipality was responsible for local social stability. Once a large-scale protest occurred, he was under pressure and had to find a way to end it as soon as possible. The giving in strategy – indicated by project relocation – could help him achieve this. When the participants involved in the protest became aware that Dalian Municipality had publicly promised to stop the operation of the Fujia PX plant, they did not have a tenable excuse to continue their protest actions. As argued by respondent 24, as the municipality had made the relocation decision and fulfilled the protesters' request, the citizens now had no legal excuse for further protest. As a result, the application of a giving in strategy made the large-scale protest lose momentum.

Proposition 1 in Chapter 3 was: local governments tend to apply a go-alone or a suppression strategy to cope with peaceful protests, whereas they tend to adopt a tension reduction or a giving in strategy to deal with violent protests. In this case, this proposition was disconfirmed and reformulated: local governments may apply a giving in strategy to cope with peaceful protests. In addition, proposition 2 was: local governments tend to apply a tension reduction or a giving in strategy to cope with large-scale protests, whereas they tend to apply a go-alone or a suppression strategy to deal with small-scale protests. This proposition was confirmed: local governments may apply a giving in strategy to deal with large-scale protests.

The explanation for the application of a tension reduction strategy

At the end of the Dalian PX case, the Fujia PX plant resumed operation secretly; this is characterized as a tension reduction strategy.

Some other strategies (such as go-alone, giving in, collaboration, or facilitation) were not appropriate for Dalian Municipality in the end. First, because of the substantial costs of removing the Fujia PX plant, it was inappropriate for it to apply a giving in strategy. The Fujia PX plant had operated in Dalian for several years. Its stoppage would have negatively reduced Dalian Municipality's tax inflows. And its relocation would be very tricky for Dalian Municipality, as it would have required large amounts of finance, as argued by respondents 26 and 27. Even worse, no actor was ready to assume such costs. Respondent 29 was doubtful about the feasibility of the relocation decision because of the lack of financial resources. Second, a go-alone strategy was not feasible for Dalian Municipality either. The municipality had already promised local citizens that the Fujia PX plant would be relocated. If it publicly refused to do so (the application of a go-alone strategy), its authority and legitimacy would be negatively damaged. This might have resulted in another round of protests initiated by local citizens against the PX plant.

In addition, it was unnecessary for Dalian Municipality to collaborate with local residents. Since the large-scale protest in Dalian, public attention on the Fujia PX plant had gradually faded over time, and Dalian Municipality had promised that the Fujia PX plant would be relocated in a concrete way. It would thus not have been easy for local citizens to find an excuse to organize another large-scale protest to oppose the plant's operation. Therefore, Dalian Municipality had a low motivation to resolve the conflict regarding the operation of the plant in a collaborative way. A facilitation strategy would have required substantial coordination work, and the leaders of Dalian Municipality might not have had a strong incentive to invest their time in doing this.

Therefore, the application of a tension reduction strategy was a better option for Dalian Municipality in the end. Two conditions are important in explaining this: the support of higher-level governments and the late stage of the Fujia PX plant. After the large-scale protest, the NDRC together with the other four ministries released a notice pointing out that the protest initiated by local residents was an event resulting from unsafe PX production. It demanded that the Fujia PX plant should operate in a safe way. This implies that the central government did not oppose the resumption of the Fujia PX plant. Dalian Municipality tended to view this as implicit support, and this contributed to its application of a tension reduction strategy. Moreover, the Fujia PX plant was in its operation stage, implying that it would be very costly for Dalian Municipality to give it up. This also contributed to its application of a tension reduction strategy.

Proposition 5 in Chapter 3 was: local governments tend to apply a tension reduction, a giving in, a collaboration, or a facilitation strategy when higher-level governments criticize debated projects, whereas they tend to adopt a go-alone or a suppression strategy if higher-level governments support debated projects. In this case, this proposition was disconfirmed and reformulated: local governments may adopt a tension reduction strategy when higher-level governments (here mainly the central government) support debated projects. In addition, proposition 4 was: local governments tend to apply a tension reduction, a giving in, a collaboration, or a facilitation strategy when debated projects are in the planning stage, whereas they tend to apply a go-alone or a suppression strategy when debated projects are in their late stage. This proposition was disconfirmed and reformulated: local governments may adopt a tension reduction strategy when debated projects are in their late stage.

Discussion

In some respects, the Dalian PX case seems to be a representative case. This is one reason for choosing this case for in-depth study. Some typical scenarios regarding the governance of conflicts concerning the planning, construction, and operation of industrial plants in China include strong alignment relationships between local industries and local governments, strong economic reliance on industrial plants for local economic development, and citizens' difficulties in stopping the construction or operation of these industries (Liu et al. 2016). These characteristics imply that local governments tend to prioritize local economic development, and they have a low willingness to accommodate the demands of citizens. The Dalian case seems to have some of these characteristics. Dalian is a traditional industrial basis in China, with many industrial plants located there. Dalian Municipality has a strong economic reliance on these industrial plants. In addition, the operation of the Fujia PX plant held the prospect of substantial economic benefits for local development in Dalian. It was pointed out by respondent 22 that the top leaders of Dalian Municipality had a close relationship with the private CEO of the Fujia PX plant. This alignment relationship might have shaped the municipality's priority along local economic interests rather than other values (such as sustainability). In short, comparison of the Dalian case with other cases may further elucidate the generic versus the unique nature of this case.

Conclusions

Two issues were addressed in this single case study. The first was about which strategies were applied by Dalian Municipality, and the second was about how to explain the application of these strategies. Regarding the first issue, five key government strategies were adopted by Dalian Municipality to govern the conflict over the operation of the Fujia PX plant: a go-alone, a suppression, a tension reduction, a giving in, and again a tension reduction strategy. In this case, a collaboration or a facilitation strategy did not emerge. To date, the Fujia PX plant is continuing to operate although the municipality promised to remove it.

The application of five government strategies in the Dalian case were explained using seven conditions: the form of protest, the scale of protest, the stage of the project, the position of the national mass media, the position of higher-level governments, the involvement of activists, and the occurrence of events. In general, four conditions are crucially important in explaining the application of, or shifts in, strategies applied by Dalian Municipality in this case:

the position of higher-level governments, the stage of the project, and the scale and the form of protest. Because of the high support from higher-level governments, the late stage of the Fujia PX plant, and the unsustained protest by citizens, Dalian Municipality finally decided to continue the operation of the PX plant.

In addition, five propositions were reformulated in this case study.

1   Proposition 1: Local governments may apply a giving in strategy to cope with peaceful protests.
2   Proposition 4: Local governments may apply a tension reduction strategy when a debated plant is in its late stage.
3   Proposition 5: Local governments may adopt a tension reduction strategy when higher-level governments show their support.
4   Proposition 6: Local governments may apply a suppression strategy to cope with the involvement of embedded activists.
5   Proposition 7: Local governments may adopt a tension reduction strategy to deal with the occurrence of unplanned events.

The above conclusions provide us with in-depth knowledge about how Chinese local governments govern environmental conflicts and why they apply various strategies.

In the following two chapters, two comparative case studies are documented. In Chapter 7, a comparative study using both the method of agreement and the method of difference is reported with the aim of analyzing how to explain similarities and differences in patterns of government strategies during environmental conflicts. In Chapter 8, the second comparative case study applying QCA is reported, studying how combinations of conditions shape the application of government strategies during environmental conflicts.

Notes

1   www.miit.gov.cn/n11293472/n11295125/n11299455/11753555.html, available on May 26, 2015.
2   http://focus.cnhubei.com/columns/columns4/201109/t1830576.shtml, available on May 26, 2015.
3   www.dlxww.com/gb/daliandaily/2009-06/22/content_2738891.htm, available on May 26, 2015.
4   www.lifeweek.com.cn/2011/0826/34723_5.shtml, available on May 26, 2015.
5   www.infzm.com/content/91197, available on May 26, 2015.
6   http://tv.people.com.cn/GB/14645/15359251.html, or www.ftchinese.com/story/0010 40134?full=y, available on May 26, 2015.
7   www.lifeweek.com.cn/2011/0826/34723_6.shtml, available on May 26, 2015.
8   www.ajj.dl.gov.cn/Simplified/NewsShow.aspx?NewsId=2815, available on May 26, 2015.
9   www.guancha.cn/local/2012_12_27_116915.shtml, available on May 26, 2015.
10 http://money.163.com/11/0929/14/7F4H8UOC00253B0H.html, available on May 26, 2015.
11 http://news.ifeng.com/mainland/detail_2011_11/01/10310346_0.shtml, available on May 26, 2015.
12 www.guancha.cn/local/2012_12_27_116915.shtml, available on May 26, 2015.

13 http://news.qq.com/a/20111230/001367.htm, available on May 26, 2015.
14 http://news.sina.com.cn/c/sd/2011-12-30/153723725020.shtml, available on May 26, 2015.
15 www.theguardian.com/environment/2012/jan/13/chinese-chemical-plant, available on May 26, 2015.
16 Information blockage would not have been an effective way for Dalian Municipality to end immediately the large-scale protest initiated by local citizens either.

7 What makes the patterns of government strategies similar and different during environmental conflicts?

Introduction

Two empirical studies have been reported in Chapters 5 and 6. They provide in-depth insights into the dynamic shifts in government strategies in environmental conflicts. In this chapter, a comparative case study of 10 cases (see Appendix 1) is reported to answer the research question: *How can the similarities and differences in the patterns of government strategies during environmental conflicts be explained?* To answer this question, both the method of agreement and the method of difference are used to make a two-level analysis: analyzing the *within-pattern (cross-case)* and the *cross-pattern level*. This chapter aims at further contributing to the explanation of the application of government strategies during environmental conflicts.

This chapter is structured as follows. The method used is elaborated in the first section, and three patterns of government strategies are then identified. In the next section, three within-pattern comparisons are made, and three cross-pattern comparisons are conducted in the following section. In the final section, conclusions are drawn.

Research method

Before elaborating the method used, we first introduce the 10 cases studied in this chapter. In Table 7.1, some general information about them is presented. More detailed information on the 10 cases is provided in Appendix 1.

In this chapter, both the method of agreement and the method of difference are employed (Collier and Collier 1991; Mill 1843; Moore 1993; Przeworski and Teune 1970; Skocpol 1979).

- *The method of agreement or the Most Different System Design (MDSD)*: The MDSD is used to study cases or situations with many differences, but with similar outcomes. The big puzzle is how to explain this similarity. Normally, the conditions with the same values across cases have important explanatory power to do this.
- *The method difference or the Most Similar System Design (MSSD)*: The MSSD is applied to study cases with many similar conditions that have

different outcomes. The question is which conditions matter to this difference. The conditions with different values across cases are important in explaining it.

Some key debates about these two methods were presented in the "Comparative case study" section in Chapter 4 (see Lieberson 1991, 1994; Ragin 1987; Savolainen 1994). I have three basic positions about the use of the two methods in this chapter.

First, I view the two methods as important analytic tools to structure case comparisons. The method of agreement is used to compare cases in a similar pattern of government strategies, whereas the method of difference is used to compare cases in different patterns of government strategies. Second, I follow this elimination logic of inference and use the two methods to identify which conditions are relatively important to explain similarities and differences regarding *the patterns of government strategies* during environmental conflicts. Third, because of their limitations, such as their failure to handle the interaction of conditions, the possibility of drawing spurious conclusions, or their unstructured analytic procedures, I do not use the two methods as appropriate approaches to identify the necessary or sufficient conditions. In fact, qualitative comparative analysis (QCA) is a better option to seek necessary or sufficient conditions. This is done in Chapter 8. In this chapter, a two-level analysis, namely, the within-pattern (or cross-case) level and the cross-pattern level, is made.

- *Within-pattern* or *cross-case analysis* means that a case is viewed as a whole, and cases in the same pattern of government strategies are compared. The method of agreement is used to identify the conditions shared by cases in the same pattern. The aim is to identify conditions that are important in explaining their similarities.
- *Cross-pattern analysis* implies that differences regarding the patterns of government strategy are the main interest. The 10 cases studied in this chapter show different patterns of government strategies (details in the section on "Within-pattern comparison using the method of agreement"). The method of difference is employed to identify which conditions matter in explaining this difference.

In Table 7.2, an overview of the relationships between the two levels of analysis and the method used is presented. Next, the patterns of government strategies are first identified and then the two levels of comparisons are made.

Identification of the three patterns of government strategies during environmental conflicts

It is necessary first to have some general understanding of the strategies adopted by local governments in the 10 cases. Government strategies in this comparative analysis are demarcated by three time anchors: before protest, during protest,

Table 7.1 Some general characteristics of the 10 cases of environmental conflict in China

Case	Location	Period	Brief description of the case
Ningbo PX	Ningbo	October 2012	When local residents realized that a PX plant would be constructed, they took to the streets to protest against it. Three protests, maximally involving about 1,000 local residents, occurred, and the local government eventually abandoned the project.
Xiamen PX	Xiamen	November 2006–December 2007	A peaceful protest, involving 8,000 to 10,000 participants, was initiated by local residents to oppose the construction of a PX plant. The State Environmental Protection Administration advised Xiamen Municipality to reconsider its decision. The planned project was finally relocated to Zhangzhou, another city in Fujian province.
Dalian PX	Dalian	August 2011–December 2012	About 12,000 local residents held a peaceful protest against the Dalian PX plant. The mayor of Dalian Municipality promised to relocate the PX plant. However, after the initial closure of the plant, operation was resumed in secret.
Kunming PX	Kunming	April–September 2013	Two protests against the construction of a PX plant occurred, initiated by local residents, both involving over 2,000 participants. The PX plant was part of a mega project. The mayor promised that the project would be cancelled, but later the planning of the project was resumed, until time of writing at least.
Pengzhou PX	Chengdu	May–October 2013	In Pengzhou, as part of a mega project including a PX plant, a refinery was constructed. Following an earthquake, local residents realized the safety risks involved and hence planned to organize a protest. The planned protest was prevented by the local government. It was reported that the PX plant was finally built secretly.

Panyu incineration power plant	Guangzhou	December 2009–April 2013	Some activists used social media to attract public attention and to pressurize local governments in Guangzhou. This resulted in a peaceful protest involving about 500 participants. Eventually, the planned waste incineration power plant was relocated.
Liulitun incineration power plant	Beijing	December 2006–January 2011	National mass media reported extensively on the debate about the Liulitun waste incineration power plant in Beijing. In response, the national government ordered local governments to reconsider the project. About 1,000 participants gathered in a peaceful protest, eventually resulting in the relocation of the plant.
Tianjingwa incineration power plant	Nanjing	September 2008–November 2011	When about a hundred local residents went to the local government to express their discontent with the construction of a waste incineration power plant, they were attacked by anonymous people. Finally, the proposed waste incineration power plant was relocated.
Songjiang incineration power plant	Shanghai	May 2012–December 2013	Local governments decided to construct a waste incineration power plant, leading to strong opposition from local residents. About 600 residents took to the streets to express their opposition. A small-scale violent confrontation occurred between the local residents and governments. In the end, the originally planned waste incineration power plant project was relocated.
Wuxi incineration power plant	Wuxi	March 2011–May 2012	Local governments built a waste incineration power plant in Wuxi to resolve the waste problem. The local residents were unaware of this until the waste incineration power plant started its trial operation. A large-scale protest involving about 10,000 local residents occurred, initiated by local citizens. Local governments promised to temporarily halt the project. Afterwards, when some residents gathered to protest, local governments used state force to disperse them, resulting in a strong violent confrontation involving about 4,000 anti-riot police officers. Finally, the constructed waste incineration power plant was dismantled.

Source: Li et al. (2016b).

Table 7.2 The relationship between the level of comparison and the methods used

Level of analysis	Method used	Main aim
Within-pattern (cross-case)	The method of agreement	Identify which conditions are important in explaining the similarities in cases in the same pattern of government strategies.
Cross-pattern	The method of difference	Identify which conditions are important in explaining the differences in cases in different patterns of government strategies.

and after protest. Government strategies applied by local governments in the 10 cases are shown in Table 7.3. Four main findings are identified and reported as follows:

- *Local governments tend to ignore the complaints of other actors at the beginning of environmental conflicts.* Table 7.3 reveals that local governments adopt mostly a go-alone strategy or a suppression strategy to advance industrial projects. This means that the doing-it-alone decision-making style dominates in the project planning or construction stage when local governments receive little pressure from outsiders (e.g., the occurrence of protests).

- *All cases are ended by various degrees of change compared to the initial government decisions.* Local governments in three cases, the Dalian case, the Kunming case, and the Pengzhou case, finally continued the debated projects. In the Ningbo case, the Tianjingwa case, the Songjiang case, the Liulitun case, and the Wuxi case, local government ultimately applied a giving in strategy, indicated by project relocation. In both the Panyu case and the Xiamen case, local governments relocated the plants, as well as engaging local citizens in formal decision-making processes.

- *Cases show different paths toward the same outcome.* Table 7.3 reveals that cases with an identical outcome take different paths toward it. Take the Liulitun case and the Ningbo case as two examples. The Liulitun case follows the pathway of go-alone–tension reduction–giving in, whereas the Ningbo case follows the pathway of tension reduction–suppression–giving in.

- *Suppression and tension reduction are the strategies applied by local governments to cope with the occurrence of protests.* Local governments in Dalian, Guangzhou, Xiamen, Kunming, Nanjing, and Beijing adopted a tension reduction strategy to deal with protests, whereas local governments in Pengzhou, Ningbo, Wuxi, and Songjiang adopted a suppression strategy.

Table 7.3 Dynamic shifts in government strategies during 10 environmental conflict cases

Case	Government strategy before protests	Government strategy during protest	Government strategy after protest
Dalian case	1 *GA*: Local mass media framed the PX plant as an important project for local economic development. 2 *SU*: An embedded activist quit his job when he questioned government decisions. 3 *TR*: After dyke damage at the Fujia PX plant, Dalian Municipality claimed that a financial investment would be contributed to avoid the occurrence of incidents again.	*GI*: The mayor of Dalian Municipality had a face-to-face talk with citizens and stated that the PX plant would be removed.	*TR*: The PX plant continued to operate secretly.
Xiamen case	1 *TR*: The key leaders of Xiamen Municipality had a face-to-face conversation with an activist, Zhao Yufen, who questioned the legitimacy of constructing a PX plant in Xiamen. 2 *GA*: The Xiamen PX plant started formal construction.	*TR*: Xiamen Municipality claimed that the PX plant would be halted and it would make a comprehensive environmental assessment for the whole Haicang region.	1 *FA*: Xiamen Municipality organized online voting and a face-to-face conversation that was broadcast live. 2 *GI*: Xiamen Municipality relocated the PX plant to Zhangzhou.
Kunming case	*SU*: One post online released by an anonymous activist questioned the legitimacy of constructing a refinery plant in Kunming. Afterwards, Kunming Municipality deleted this post, and relevant words were screened.	*TR*: The mayor of Kunming Municipality claimed that the Kunming refinery plant would not be constructed as long as it was opposed by most citizens in Kunming.	*TR*: Kunming Municipality organized local residents to review the environmental impact report on the refinery project. The project was finally progressed secretly.

continued

Table 7.3 Continued

Case	Government strategy before protests	Government strategy during protest	Government strategy after protest
Ningbo case	**TR**: Government officials in Zhenhai, Ningbo, had a face-to-face conversation with local residents and promised that they would be resettled in a new location.	**SU**: Local governments in Ningbo used state force to disperse local residents, and some of them were arrested.	**GI**: The Ningbo PX plant was cancelled.
Pengzhou case	*1* **TR**: After an earthquake in Sichuan, the Pengzhou refinery plant project was temporarily halted. *2* **GA**: The China Earthquake Administration (CEA) concluded that the refinery plant met the technical requirement established by national laws, and local governments in Pengzhou further advanced the refinery plant. *3* **TR**: Another earthquake occurred, and the refinery project was temporarily stopped.	**SU**: Local governments in Pengzhou took prohibitive measures to prevent the occurrence of a protest planned by local residents.	*1* **TR**: Local governments in Pengzhou organized local residents to visit the Pengzhou refinery plant. *2* **TR**: The Pengzhou refinery plant started formal operation and no information about the Pengzhou PX plant was released.
Panyu case	*1* **GA**: Local governments in Guangzhou advanced the construction of the Panyu waste incineration power plant without informing local residents. *2* **SU**: Some activists were formally requested to attend the local police department in Guangzhou.	**TR**: Local governments in Guangzhou publicly stated that the Panyu waste incineration power plant would not be advanced.	*1* **CO**: Local governments in Guangzhou organized two public symposiums. And they relocated the waste incineration power plant and compensated local residents living near its new location. *2* **FA**: Waste classification was established as a policy to resolve the urban waste problem.

Case			
Liulitun case	**GA**: Local governments advanced the Liulitun waste incineration power plant although local residents visited different government agencies to express their opposition.	**TR**: The key leaders of local governments in Beijing had a face-to-face conversation with some activists in Liulitun.	**GI**: Local governments in Beijing stated that the Liulitun waste incineration power plant was relocated unilaterally.
Tianjingwa case	**GA**: Local governments in Nanjing advanced the construction of the Tianjingwa waste incineration power plant without informing local citizens.	**TR**: When local residents in Tianjingwa went to the Jiangsu Provincial Environmental Protection Department (JPEPD), they were suppressed by persons from the JPEPD. Government officials from Nanjing Municipality went there to persuade residents to return home.	**1 GA**: Local governments in Nanjing further advanced the Tianjingwa waste incinerator. **2 GI**: Local governments in Nanjing relocated the Tianjingwa waste incinerator unilaterally.
Songjiang case	**1 GA**: Local governments did not resolve the odour problem resulting from waste landfill. **2 GA**: Local governments decided to construct a waste incineration power plant despite local residents' disagreement.	**SU**: Local residents and local government officials in Songjiang had a small-scale violent confrontation.	**1 TR**: Local governments in Songjiang apologized to local residents. **2 GI**: The Songjiang waste incineration power plant was relocated unilaterally.
Wuxi case	**1 GA**: Local governments in Wuxi lied to local residents in order to advance the construction of the Wuxi waste incinerator. **2 TR**: Local government claimed that the waste incineration power plant would not be advanced.	**SU**: Local residents were repressed when they attempted to express their opposition to the operation of the waste incineration power plant.	**GI**: The constructed Wuxi waste incineration power plant was abandoned.

Note
GA = go-alone strategy, SU = suppression strategy, TR = tension reduction strategy, GI = giving in strategy, CO = collaboration strategy, FA = facilitation strategy.

Before the within-pattern and cross-pattern comparisons are made, it is necessary to identify patterns. In general, three different substantive outcomes regarding the debated industrial plants are identified: project abandonment, project continuation, and project relocation. Three different patterns of government strategies are thus established in this chapter (see Table 7.4).

The first pattern is indicated by the final decision to cancel the project. The Ningbo case and the Wuxi case are such examples. The second pattern is indicated by the final decision to continue the project. The Dalian case, the Kunming case, and the Pengzhou case are such examples. The third pattern is indicated by the final decision to relocate the project. Five cases – Tianjingwa, Xiamen, Songjiang, Liulitun, and Panyu – are such examples. Now that the three patterns of government strategies have been identified, their similarities and differences can be explained. This is done in the following two sections.

Within-pattern comparison using the method of agreement

In this section, the method of agreement is used to explain the similarities in the cases in the same pattern of government strategies. If conditions co-vary with outcomes, they are important in explaining the similarities of the same pattern of government strategies. Specifically, if conditions are similar, they in principle explain outcomes. The three patterns of government strategies are compared in the following three sections.

Understanding the similarities in the two cases in pattern 1

The first pattern of government strategies covers two cases: the Ningbo case and the Wuxi case. At the beginning of the Ningbo case, local governments applied a tension reduction strategy with the aim of avoiding protests by local residents. In the Wuxi case, local governments did not inform local residents that a waste incineration power plant would be constructed near their communities. During both cases, local governments used state force to repress local residents. Eventually, the disputed projects in both cases were abandoned by the local governments. One of the crucial similarities between the two cases is that both local governments eventually applied a giving in strategy, indicated by project abandonment. The values of the conditions in the two cases are presented in Table 7.5.

In broad terms, five of the seven conditions show similar values in both cases: violent protest, silence of higher-level governments, silence of national mass media, absence of activists, and absence of events. These factors may be important in explaining why both projects were cancelled. Below, the influence of each of the conditions is discussed.

- *Condition 1: The form of protest*: Violent confrontations occurred in both cases. Violent protest in Ningbo came as a surprise to local governments. A small group of local residents in Zhenhai first asked to be put on the list for

Table 7.4 Three patterns of government strategies in 10 cases

	Case	Government strategy before protest	Government strategy during protest	Government strategy after protest
Pattern 1	Ningbo case	Tension reduction	Suppression	Giving in (indicated by project abandonment)
	Wuxi case	Go-alone–tension reduction	Suppression	Giving in (indicated by project abandonment)
Pattern 2	Dalian case	Go-alone–suppression–tension reduction	Giving in	Tension reduction (indicated by project continuation)
	Kunming case	Suppression	Tension reduction	Tension reduction (indicated by project continuation)
	Pengzhou case	Tension reduction–go-alone–tension reduction	Suppression	Tension reduction (indicated by project continuation)
Pattern 3	Tianjingwa case	Go-alone	Tension reduction	Go-alone–giving in (indicated by project relocation)
	Xiamen case	Tension reduction–go-alone	Tension reduction	Facilitation–giving in (indicated by project abandonment)
	Songjiang case	Go-alone–go-alone	Suppression	Tension reduction–giving in (indicated by project relocation)
	Liulitun case	Go-alone	Tension reduction	Giving in (indicated by project relocation)
	Panyu case	Go-alone–suppression	Tension reduction	Collaboration and facilitation (indicated by project relocation)

Table 7.5 The values of the conditions in the two cases in pattern 1

Case	Conditions						
	Form of protest	*Scale of protest*	*Higher-level governments*	*National mass media*	*Stage of projects*	*Activists*	*Events*
Ningbo case	*violence*	1,000	*silence*	*silence*	planning	*no*	*no*
Wuxi case	*violence*	10,000	*silence*	*silence*	before operation	*no*	*no*

Note
Italics indicate conditions that are important in explaining pattern similarities.

the new resettlement project proposed by local governments, which the latter promised to do. However, many citizens in the center of Ningbo City afterwards were surprised that a PX plant would be constructed, following which a protest occurred. During the protest, over 100 participants threw bricks and water bottles at police officers, and even attacked a police car.[1] Fifty-one participants were detained, and 13 of them were finally sentenced as criminals.[2] In this case, the protest initiated by local citizens lasted four days. When local governments realized that the protests were significantly undermining regime legitimacy, they finally decided to abandon the proposed PX project. In the Wuxi case, when some female senior citizens went to the town government to express their opposition to the Wuxi waste incineration power plant, some of them were arrested by local police officers. A strong violent confrontation occurred afterwards.[3] Over 100 participants were injured, and tens of them were arrested.[4] In this case, the use of state force by local governments in Wuxi seems to have been their last resort to end local residents' opposition in order to realize the operation of the constructed waste incineration power plant. However, the use of state repression resulted in casualties. Local governments in Wuxi thus were in a morally disadvantaged position. Regime legitimacy was undermined, and the tensions between local governments and local residents escalated. Finally, local governments rejected the waste incineration power plant although it had already been constructed. To sum up, violent confrontations occurred in both cases. Their occurrence negatively undermined state legitimacy, and it became politically risky for local governments to keep advancing the debated projects. Abandoning the projects was a way out for them. Therefore, the occurrence of violent protests was an important condition in explaining the cancellation of the projects in the two cases.

- *Condition 2: The scale of protest*: The scale of protest in the two cases was different. In the Wuxi case, over 10,000 local residents assembled together to express their opposition to the operation of the waste incineration power plant.[5] One senior government official became their hostage. Local governments afterwards claimed that the plant would be temporarily stopped and a commission dispatched by the Ministry of Environmental Protection (MEP) would review this project. Another protest also occurred in this case, but the number of protesters was unknown. However, it was reported about 4,000 anti-riot police officers were dispatched to handle the protest.[6] In the Ningbo case, the scale of protest was not that large. Over 1,000 local residents participated in the demonstration.[7] In conclusion, the scale of protest in the two cases varied. This implies that the scale of protest was not a crucial condition in explaining the cancellation of the projects in the two cases.

- *Condition 3: The position of higher-level governments*: Higher-level governments in neither case publicly expressed support for the debated projects after the protests had occurred. Prior to the violent confrontation in the Wuxi case, the MEP had dispatched a commission to review the Wuxi waste incineration power plant and concluded that it could be operated provided its facilities were upgraded.[8] After the violent confrontation, the national government did not express its position. In the Ningbo case, the national government kept silent throughout. To sum up, the higher-level governments in the two cases kept silent after the protests, implying that they were reticent about showing their positions. This was understandable. If they had publicly supported the debated projects after the violent confrontations, they would have been criticized by the citizens. This would have significantly undermined regime legitimacy. Consequently, the absence of support from higher-level governments seems to have been an important condition contributing to the cancellation of the projects in the two cases.

- *Condition 4: The position of the national mass media*: No national mass media were involved in either of the two cases. In the Wuxi case, all the national mass media kept silent, and all information about the violent confrontation between local governments and local citizens was censored.[9] The national mass media did not publicly report the Ningbo case either. After the occurrence of the violent confrontations in the two cases, when the national mass media did not publicly show their opposition to the existing strategies applied by the two local governments, the latter tended to take action to avoid triggering public attention on the debated plants. Project abandonment was a possible option for them to achieve this. Therefore, it can be concluded that the absence of support from the national mass media contributed to the cancellation of the debated projects in the two cases.

- *Condition 5: The stage of the debated projects*: The projects in the two cases were at different stages. Previously, the citizens in Zhenhai had been heavily reliant economically on the Ningbo refinery. They thus tended not to complain about its negative effect on the local environment and public health. However, the new generations were less reliant on the refinery.[10] When it was decided to construct a new PX plant nearby, local residents were concerned about the quality of the local environment and asked for economic compensation from local governments (resettlement in this case).[11] Meanwhile, most of the economic profit from the refinery went to the China National Petroleum Corporation. Consequently, the local governments in Ningbo did not have a strong incentive to align themselves with the refinery to advance the PX plant.[12] When they realized that local citizens were strongly against the PX plant in its planning stage, rejecting it became an attractive option. In the Wuxi case, the waste incineration power plant was already undergoing trial operation and testing.[13] Consequently, cancellation would be potentially costly for the local governments. Moreover, it was an important facility for them to dispose of urban waste in Wuxi. Local governments in Wuxi thus had a strong incentive to advance this plant. Initially, the local governments did not inform local residents, who did not know that a waste incineration power plant would be constructed nearby until it started its trial operation.[14] In summary, the two debated projects were in different stages and therefore entailed different benefits and costs for the local governments involved. Compared to local governments in Ningbo, those in

Wuxi had a relatively stronger incentive to advance their project. At the end of the Wuxi case, local governments used state force to repress citizens in order to realize the operation of the waste incineration power plant, even though this might incur high political risks for them. In the Ningbo case, however, it seems that local governments gave up the project comparatively easily. Overall, the stage of the projects was not a critical condition in explaining the cancellation of the projects in the two cases.

- *Condition 6: The involvement of activists*: Activists were not involved in either case. Cai (2010) concluded that the absence of activists makes violent confrontation more likely because it is more difficult for an environment of negotiation and dialogue between local governments and local residents to emerge. In the Ningbo case, local residents spontaneously took to the streets to oppose the construction of a PX plant.[15] Neither coordinators nor leaders were present. The Wuxi case lasted over a year, from January 2011 to May 2012. No activists were present either. One possible explanation may be the type of protesting community (Cai 2010). The protesters were mostly from different communities. It was difficult for them to have a strong leader to coordinate their actions. To sum up, there were no activists in either case. This implies that the absence of activists may be an important condition in explaining the similarity of the two cases.
- *Condition 7: The occurrence of events*: There were no eye-catching events in either case. In this respect, the value of this condition in the two cases is the same. Therefore, the absence of events may be important in explaining the similarity of the two cases.

To sum up, the occurrence of strong violent confrontations between local governments and local residents, and the absence of support from higher-level governments, the absence of activists, the absence of events, and the absence of support from the national mass media are the five relatively crucial conditions in explaining the cancellation of the projects in the two cases. The other two conditions, namely, the scale of protest and the stage of the projects, do not have substantial influence.

Understanding the similarities in the three cases in pattern 2

Three cases are covered by the second pattern of government strategies: the Dalian, the Kunming, and the Pengzhou case. At the beginning of the three cases, local governments applied a go-alone or a suppression strategy to advance the construction of the three projects. During the cases, Dalian Municipality and Kunming Municipality, respectively, adopted a giving in and a tension reduction strategy to cope with the protests initiated by local citizens. Unlike these two cases, local governments in Pengzhou took some preventive actions to impede the occurrence of protests planned by local citizens. Finally, local governments in the three cases continued the debated projects. In general, the three cases share a crucial similarity: all local governments eventually adopted a tension reduction strategy, indicated by project continuation in secret.

The values of the seven conditions in the three cases are presented in Table 7.6. The three cases share a condition with the same value: peaceful protest. This implies that the occurrence of peaceful protest may be important in explaining the continuation of the projects in the three cases.

- *Condition 1: The form of protest*: No violent confrontations occurred in any of the three cases. When protests were peaceful, they did not radically endanger social order and threaten regime legitimacy. Local governments then were not under much pressure to treat seriously the disagreements expressed by local citizens. They hence tended to further advance the debated projects. In conclusion, the occurrence of peaceful protests seemed to contribute to the continuation of the projects in the three cases.
- *Condition 2: The scale of protest*: The scale of protest in the three cases varied. A large-scale protest involving about 12,000 participants occurred in Dalian.[16] In the Kunming case, the two protests involved about 2,000 participants.[17,18] One similarity between two cases was that the mayors of both Dalian and Kunming Municipality had a face-to-face talk with the protest participants.[19] In the Pengzhou case, a planned large-scale protest did not occur. Chengdu Municipality replaced Saturday and Sunday with Monday and Tuesday. Citizens in Chengdu had to work on Saturday and Sunday, as if those days were a Monday and Tuesday. Also, all students were required to study at their schools to avoid their involvement in the planned demonstration.[20] Some sites in the city center were closed to visitors, and many police officers were on guard there.[21] Because of these preventive actions, the planned large-scale protest did not occur. To sum up, the scale of protest was different in the three cases, implying that scale of protest was not crucial in explaining the continuation of the projects in the three cases.
- *Condition 3: The position of higher-level governments*: Higher-level governments all supported the debated projects in the three cases. In the Dalian PX case, Dalian was an important industrial base in China. Local governments in Dalian had a strong economic reliance on industrial plants. Before its formal approval by the Liaoning Provincial Environmental Protection Department, the PX plant had started formal operation.[22] In addition, the Dalian PX plant was established as a key industrial project by the National Development and Reform Commission (NDRC). After the occurrence of large-scale protest in Dalian, the NDRC, together with four other ministries, framed the occurrence of the protest as the result of unsafe PX production.[23] In the Pengzhou case, after the occurrence of an earthquake, the State Earthquake Administration assessed the refinery and agreed with its further advancement.[24] In the Kunming case, Kunming Municipality was dissatisfied with the fact that the whole province had to import lots of refined oil from other regions of China. It attempted to construct a mega refinery plant in Kunming, and it made substantial efforts to get this project approved.[25] Both the refinery plant and the PX plant in Kunming were key projects established by it.[26] Thus, Kunming Municipality had a strong incentive to

Table 7.6 The values of the conditions in the three cases in pattern 2

Case	Conditions						
	Form of protest	*Scale of protest*	*Higher-level governments*	National media	*Stage of project*	*Activists*	*Events*
Dalian case	*peaceful*	12,000	*support*	silence	*operation*	embedded	unplanned (explosion)
Kunming case	*peaceful*	2,000	*support*	support	*planning*	unembedded	planned (China-South Asia Expo)
Pengzhou case	*peaceful*	no	*support*	support	*trial operation*	no	unplanned (earthquake) and planned (Fortune Global Forum)

Note
Italics indicate conditions that are important in explaining pattern similarities.

realize the construction of these two projects. In conclusion, the higher-level governments all supported the advancement of the three debated projects, and this contributed to the continuation of the projects at the end of the three cases.

- *Condition 4: The position of the national mass media*: The national mass media adopted various positions in the three cases. In both the Kunming and the Pengzhou case, the most authoritative national television – China Central Television (CCTV) – and the most authoritative mouthpiece of the Chinese Communist Party (CCP) – *Renmin Daily* – publicly stated that the refinery projects were necessary for national development and PX was not as toxic as citizens thought.[27] In the Dalian case, the reporters dispatched by CCTV attempted to investigate the PX plant, but they were not allowed to enter it. The investigation result was not broadcast afterwards on CCTV.[28] In summary, the national mass media in the three cases did not have the same position. This condition, the position of the national mass media, thus is not important in explaining the continuation of the projects in the three cases.

- *Condition 5: The stage of the projects*: The projects in the three cases were in different stages. The PX plant had operated in Dalian for several years. Removing it would require large amounts of finance (respondents 26, 27, 28, and 29), and Dalian Municipality would not like to assume that responsibility (respondent 29). Furthermore, the PX plant contributed substantially to local GDP, implying that the Dalian Municipality would prefer to continue rather than remove or reject it. In the Pengzhou case, the refinery project was already constructed. To remove or abandon it would have been very costly for the local governments, and citizens' opposition to the Pengzhou refinery plant was untenable for the local governments. Most local citizens were also worried about the potentially negative influence of the PX plant on their health. In the Kunming refinery project case, although it was in its planning stage, an oil pipeline from Myanmar had been constructed in order to deliver the crude oil to the plant.[29] Abandoning or removing the refinery plant would be very costly for the local governments. In short, the stages of the debated projects in the three cases varied; however, abandoning or relocating them implied high costs for the local governments in the three cases. Consequently, the local governments tended to continue them. Thus, it can be concluded that, although the stage of the debated projects was different, abandonment of the projects implied high costs for local governments, and the (perceived) high costs were important in explaining the continuation of the projects in the three cases.

- *Condition 6: The involvement of activists*: The three cases had different values regarding this condition. In the Pengzhou case, no activists were involved. Activists were involved in the Dalian and the Kunming case. When an unembedded activist in the Kunming case questioned the government decision to build the Kunming PX project, relevant information afterwards was censored.[30] Furthermore, some local environmental NGOs asked for information disclosure; however, this was ignored by the local governments.[31] In the Dalian case, when a reporter from Dalian Television asked for information disclosure about the PX plant, he soon quit his job, reportedly under pressure.[32] In the Dalian and the Kunming case, embedded and unembedded activists attempted to challenge government decisions using social media. Both local governments adopted a suppression strategy to cope with this. In brief, the values of the condition, the involvement of activists, were different in the three cases. Although the involvement of activists might play a role in influencing shifts in government strategies at a certain time, it was not crucially important in explaining the continuation of the projects in these three cases.

- *Condition 7: The occurrence of events*: There were events, planned or unplanned, in the three cases. In the Dalian case, after the occurrence of the fire at the CNPC site, local residents speculated in panic about the dreadful impacts if the Fujia PX project were to have some kind of incident too, following which they organized a large-scale protest. In the Pengzhou case, an earthquake triggered the worries of local residents about the negative influence of the refinery plant on their safety.[33] They afterwards planned to initiate a large-scale protest. After the protest in both the Dalian and the Pengzhou case, local governments applied a tension reduction strategy. Thus, the causal relationships between the occurrence of events and the application of a government strategy could be established through an intervening variable, the occurrence of protests. Furthermore, the China-South Asian Expo was hosted in Kunming and the Fortune Global Forum was hosted in Chengdu. Both local governments took some preventive actions (a real-name registration system for buying T-shirts and printing posters meant that people had to register their names with the sellers using their ID) to avoid the occurrence of protests before and during their hosting of these events.[34] In conclusion, events occurred in the three cases. They did influence the application of government strategy at a certain point in time; however, their influence on the continuation of the projects in the three cases was limited. Events could attract substantial attention in the short term, and this enabled pressure to be exerted on local governments to change their decisions. However, their influence disappeared gradually. Local governments eventually tended to stick to their previous decisions. Therefore, it can be concluded that, although events occurred in all three cases and the values of this condition were similar in this respect, the occurrence of events is not a crucially important condition in explaining the continuation of the projects at the end of the three cases.

To sum up, three conditions – the absence of violent protest, support from higher-level governments, and perceived costs of the project[35] – are important in explaining project continuation. The other four conditions, the scale of protest, the position of the national mass media, the involvement of activists, and the occurrence of events, do not contribute to its explanation.

Understanding the similarities in the five cases in pattern 3

Five cases are covered in the third pattern of government strategies: the Songjiang, the Tianjingwa, the Panyu, the Xiamen, and the Liulitun case. At the beginning of the five cases, local governments mostly applied a go-alone or a suppression strategy with a strong intention to advance the construction of the proposed projects. During the cases, local governments in four of these five cases (Panyu, Liulitun, Tianjingwa, and Xiamen) applied a tension reduction strategy to deal with the protests, and local government in the Songjiang case adopted a suppression strategy to deal with the protest. At the end of the five cases, the local governments all decided to relocate the debated projects; this was one crucially important similarity that all five cases shared.

In Table 7.7, the values of the seven conditions in the five cases in pattern 3 are presented. In general, the five cases have one condition in common: the early stage of the projects, which may be an important condition in explaining the relocation of the five projects at the end of the five cases. Below, I analyze how each condition contributes to the explanation more in depth.

- *Condition 1: The form of protest*: The form of protest in the five cases was approximately the same. Peaceful protests occurred in three of the five cases: the Liulitun, the Panyu, and the Xiamen case. No strong violent confrontation occurred in the Songjiang case either. When local residents expressed their opposition to the waste incineration power plant, small-scale unrest occurred because local residents were out of control.[36] This came as a surprise to both local governments and local residents. Different from these four cases, violent confrontation occurred in the Tianjingwa case. When local residents went to the Jiangsu Provincial Environmental Protection Department (JPEPD) to express their opposition to the construction of the Tianjingwa waste incineration power plant, some individuals came out of the government building, hustling and even hitting local residents.[37] Although this was a violent confrontation, this condition did not have a fundamentally different score compared with the other four cases. Three reasons justify this. First, it was the JPEPD that used state force to repress local residents. This might not represent the position of local governments in Nanjing. Second, because of the lack of media attention, the state repression by JPEPD did not become well known around China. Thus, the protest put little pressure on local governments in Nanjing. Third, the scale of protest was small, implying that state legitimacy was not seriously undermined. In summary, the values regarding the condition, the form of protest, were more or less the same in the five cases. Thus, it might be argued that this condition matters in explaining the relocation of the projects in the five cases.
- *Condition 2: The scale of protest*: The scale of protest in the five cases was different. Small-scale protests occurred in four cases: the Songjiang, the Tianjingwa, the Panyu, and the Liulitun case. About 600 local citizens

Table 7.7 The values of the conditions in the five cases in pattern 3

Case	Condition							
	Form of protest	Scale of protest	Higher-level governments	National mass media	Stage of project	Activists	Events
Songjiang case	peaceful	600	silence	silence	early (planning)	yes (unembedded)	no
Tianjingwa case	violent	100	contradiction	silence	early (planning)	no	yes (Youth Olympic Games)
Panyu case	peaceful	500	silence	opposition	early (planning)	yes (unembedded)	yes (Asian Games)
Liulitun case	peaceful	1,000	opposition	opposition	early (planning)	yes (unembedded)	yes (Olympic Games)
Xiamen case	peaceful	8,000–10,000	opposition	opposition	early (construction)	yes (embedded)	yes (NPPCC)

Note
Italics indicate conditions that are important in explaining pattern similarities.

expressed their opposition to the construction of the waste incineration power plant in Songjiang.[38] In the Liulitun case, about 1,000 participants went to the State Environmental Protection Administration (SEPA) to express their opposition to the construction of the Liulitun waste incineration power plant.[39] In the Panyu case, hundreds of local residents went to Guangzhou Municipality to show their opposition to the construction of the Panyu waste incineration power plant.[40] In the Tianjingwa case, about 100 local residents assembled in front of the JPEPD, demanding a conversation with its key leaders.[41] Different from these four cases, a large-scale protest involving between 8,000 and 10,000 local residents occurred in the Xiamen case.[42] In conclusion, the five cases had different values regarding the condition, the scale of protests. It thus was not an important condition in explaining the relocation of the projects in the five cases.

- *Condition 3: The position of higher-level governments*: Higher-level governments did not publicly support the debated projects in the five cases. This was what clearly happened in the Songjiang and the Panyu cases. After the occurrence of a large-scale protest in Xiamen, SEPA advised Xiamen Municipality to make an environmental impact assessment (EIA) for the whole Haicang region. Xiamen Municipality later followed this advice. In the Liulitun case, SEPA ordered Beijing Municipality to postpone the Liulitun waste incineration power plant.[43] Beijing Municipality, however, refused to do so. Because of its special geographical location, Beijing Municipality had more opportunities to negotiate with the central government, according to respondent 11. Furthermore, it may not have treated SEPA's decision seriously given the relatively weak administrative status of the latter in the State Council. Nevertheless, the MEP publicly showed its opposition to the Liulitun waste incineration power plant for a second time, as reported by respondent 11. This respondent also maintained that the deputy minister of SEPA, Pan Yue, attempted to ambitiously promote SEPA's administrative status in the national government. He used the Liulitun case to institutionalize environmental values in China. In the Tianjingwa case, the position of higher-level governments was complicated. The JPEPD strongly supported the waste incineration power project, and it even used state force to repress local citizens. The MEP supported the continuation of the plant when it received the application for an administrative review from local residents.[44] Afterwards, the Ministry of Housing and Urban-Rural Development (MHURD) investigated this case but did not reveal its position.[45] Given these different positions, it cannot be concluded that the support of higher-level governments for the Tianjingwa waste incineration power plant project was very strong. All in all, the higher-level governments more or less did not strongly support the debated projects in the five cases, implying that this was an important condition in explaining the relocation of the projects in the five cases.

- *Condition 4: The position of the national mass media*: The national mass media did not publicly support the debated projects in the five cases. No national mass media reported the Tianjingwa case or the Songjiang case. The *Renmin Daily* commented at the end of the Xiamen case that removing the PX plant was the best option for Xiamen Municipality.[46] Its opposition might have been the last straw that undermined the confidence of Xiamen Municipality to advance the PX project. In the Liulitun case, the debates concerning the construction of the Liulitun waste incineration power plant were reported by a national mass medium, CCTV, which commented that the project's EIA was questionable.[47] In the Panyu case, the national mass medium, CCTV, reported the conflict regarding the construction of the Panyu waste incineration power plant, and the invited experts on the television program advised local governments in Guangzhou to facilitate public participation.[48] In short, the national mass media did not publicly show their support for the five projects. This implies that the position of the national mass media was an important condition that contributed to the relocation of the projects in the five cases.

- *Condition 5: The stage of the debated projects*: All the debated projects were in their early stage. Four of them, the Songjiang, the Tianjingwa, the Panyu, and the Liulitun waste incineration power plant were in their planning stage. If they were opposed by local residents, local governments might tend to relocate them given the low cost of doing this. The Xiamen PX plant had just started construction. Removing it was not that costly for local governments either. To sum up, the five projects were all in their early stages, and local governments finally relocated them. This means that the early stage of the debated project and the associated low perceived costs for local governments to adapt initial plans might be an important condition in explaining the relocation of the projects in the five cases.

- *Condition 6: The involvement of activists*: The five cases did not share the same value regarding this condition. When a famous scientist, Zhao Yufen, publicly questioned the government decision to build a PX plant in Xiamen, the key leaders of Xiamen Municipality had a face-to-face conversation with her.[49] However, the project was further advanced afterwards. Zhao was an academically and politically influential person. Xiamen Municipality thus had to treat her opinions seriously. Nevertheless, she was not powerful enough to change government decisions. In the Liulitun case, some retired workers turned activists continuously visited diverse government agencies in Beijing to express their opposition to the construction of the Liulitun waste incinerator. Furthermore, respondent 11, an activist in this case, claimed that activists consulted scientists, lawyers, and journalists about waste incineration, equipping themselves with the evidence-based knowledge to argue with officials. This case lasted about three years (2006–2009), and some activists established a good personal relationship with officials, according to respondent 11. In the Songjiang case, several unembedded activists personally met government officials of the Songjiang district government and reached some agreements with the aim of arranging a peaceful petition action.[50] This seems to be a negotiation between unembedded activists and local governments. In the Panyu case, several young unembedded activists strategically expressed their disagreement with the construction of the Panyu waste incineration power plant (according to respondents 17, 18, 19, and 20). Like activists in the Liulitun case, they collected information and evidence about waste incineration. In particular, they conducted an on-site investigation into the potential harm of waste incineration on the local environment and human health (according to respondent 17). They argued and debated with government officials using evidence-based information. In the Tianjingwa case, no activists were involved. This case occurred in 2009, when social media (especially the microblog) were not that popular around China. This may have hindered the organization, mobilization, and coordination of collective actions, according to respondent 5. In addition, respondent 1 claimed that the Tianjingwa region was relatively under-developed. This might imply that local residents there did not have the necessary skills and capabilities to smartly and strategically interact with officials. In conclusion, the five cases did not share the same value regarding the condition, the involvement of activists. Although the involvement of activists in some cases really played a role in influencing the shifts in government strategies at certain times, it was not crucial in explaining the relocation of the projects in the five cases.

- *Condition 7: The occurrence of events*: The five cases did not share the same value regarding this condition. There were no eye-catching events in the Songjiang case. In the Panyu and the Liulitun case, both local governments adopted a tension reduction strategy before respectively hosting the Asian Games in Guangzhou[51] and the Olympic Games in Beijing.[52] The situations in the Xiamen and the Tianjingwa case were different from what happened in the Panyu and the Liulitun case. No planned events occurred in Xiamen, but one of the most important national political events, the National People's Political Consultative Conference (NPPCC), occurred in Beijing.[53] Zhao Yufen, together with another 105 NPPCC representatives, jointly submitted a proposal to oppose the construction of the PX plant.[54] Afterwards, the Xiamen case became known around China, and this triggered the occurrence of a large-scale protest. There were no events in the Tianjingwa case either. However, a planned event, the Youth Olympic Games, was to be hosted in Nanjing. Local governments had to do some preparatory work beforehand. The construction of a waste incineration power plant was one such effort for them to maintain the city landscape.[55] Local governments in Nanjing consequently did not necessarily have to stick to their earlier plan, which was opposed by local citizens. To sum up, the five cases did not share the same value regarding the condition, the occurrence of events. It should be noted that the occurrence of events might matter in explaining shifts in government strategies at certain times, but it was not an important condition in explaining the relocation of the projects in the five cases.

In conclusion, the four conditions, early stage of the projects, the absence of strong support from higher-level governments and national mass media, and the absence of strong violent protests, are important in explaining the similarity of the third pattern of government strategy – project relocation. The other three conditions, namely, the scale of protests, the involvement of activists, and the occurrence of events, are not so important in explaining this similarity.

Toward an explanation of the three patterns of government strategies during environmental conflicts

Based on the above within-pattern comparisons, three explanations of the three patterns of government strategies are posited below:

- Strong violent protests, the absence of strong support from higher-level governments, the absence of strong support from the national mass media, the absence of activists, and the absence of events are important in explaining the first pattern of government strategies resulting in project cancellation.
- The absence of strong violent protests, the late stage of the projects (or the high costs of project relocation or cancellation), and support from higher-level governments are crucially important in explaining the second pattern of government strategies leading to project advancement.
- The early stage of the debated projects, the absence of strong support from higher-level governments and the national mass media, and the absence of strong violent protests are important in explaining the third pattern of government strategies resulting in project relocation.

Cross-pattern comparison using the method of difference

In this section, the differences in the three patterns of government strategies are explained using the method of difference. The interest is to answer what differentiates the patterns of government strategies. The values of the seven conditions for the 10 cases are shown in Table 7.8. The three pair-wise cross-pattern comparisons are reported respectively in the following three subsections.

The pair-wise comparison of the cases in pattern 1 and pattern 3

A crucial difference between the cases in pattern 1 and pattern 3 is that local governments in the former abandoned the projects, whereas those in the latter relocated them. The question that must be answered is which conditions matter to this difference? At first sight, the data in Table 7.8 seem to indicate that no single condition contributes to the explanation of this.

However, there is a crucial condition that explains this difference on closer examination: the form of protest. Strong violent confrontations occurred in both the Ningbo case and the Wuxi case. These occurrences made the two cases finally become a sensitive political issue. Political stability then was the top

Table 7.8 The values of the conditions in the 10 cases

Pattern	Case	Form of protest	Scale of protest	Higher-level governments	National mass media	Stage of project	Activists	Events	Outcome
1	NB	violent	1,000	silence	silence	early stage	no	no	project cancellation
	WX	violent	10,000	silence	silence	final stage	no	no	
2	DL	peaceful	12,000	support	silence	final stage	yes	yes	project continuation
	KM	peaceful	2,000	support	support	early stage (high costs)	yes	no	
	PZ	peaceful	0	support	support	final stage	no	yes	
3	PY	peaceful	500	silence	opposition	early stage	yes	yes	project relocation
	LLT	peaceful	1,000	opposition	opposition	early stage	yes	yes	
	TJW	violent	100	contradiction	silence	early stage	no	yes	
	SJ	peaceful	600	silence	silence	early stage	yes	no	
	XM	peaceful	8,000–10,000	opposition	opposition	early stage	yes	yes	

Source: Li et al. (2017).

Notes

NB = Ningbo case, WX = Wuxi case, XM = Xiamen case, DL= Dalian case, KM = Kunming case, PZ = Pengzhou case, PY = Panyu case, LLT = Liulitun case, TJW = Tianjingwa case, SJ = Songjiang case.

priority for local governments, which tended to take action to eliminate the potentially negative influence of the protests on state legitimacy. Project cancellation seems to be an option to achieve this. Different from the two cases in pattern 1, the protests in the five cases in pattern 3 were mostly peaceful. One exception was the Tianjingwa case. In this case, a small-scale protest was suppressed by the JPEPD. However, its influence on the relocation of the Tianjingwa waste incineration power plant by Nanjing Municipality was limited (see above). In conclusion, the protests in the five cases in pattern 3 did not seriously endanger social stability and undermine state legitimacy. The demands of local residents were kept along the lines of protecting legitimate rights to health, and an attempt was made to avoid escalating tensions with local governments.

In conclusion, the form of protest matters in explaining the difference between the cases in pattern 1 and pattern 3.

The pair-wise comparison of the cases in pattern 1 and pattern 2

A crucial difference in the cases in pattern 1 and pattern 2 is that local governments in the former abandoned the projects, whereas the latter continued them. The puzzle to be solved is which conditions are important in explaining this difference.

Table 7.8 shows that one condition may be important in explaining this difference: the form of protest. Violent protests occurred in the two cases in pattern 1, namely, the Ningbo case and the Wuxi case. Peaceful protests occurred in the three cases in pattern 2: the Pengzhou case, the Kunming case, and the Dalian case. The occurrence of violent protests tends to negatively damage the relations between government and citizens. Local governments then are likely to make compromises to remedy this. Nevertheless, the influence of peaceful protests on social order or state legitimacy was limited. Local governments therefore tended to stick to their original decisions. The position of higher-level governments is another crucial condition that explains the difference between the cases in pattern 1 and pattern 2. For the three cases in pattern 2, as argued above, higher-level governments generally supported the continuation of the three debated projects, namely, the Pengzhou refinery project, the Kunming refinery project, and the Dalian PX plant. Local governments as a result tended to continue them. Contrarily, as shown in the section on "Understanding the similarities in the two cases in pattern 1" above, higher-level governments did not publicly show their support for the Ningbo PX plant and the Wuxi waste incineration power plant after the occurrence of protests. The local governments thus did not have sufficient confidence to further advance them.

To sum up, the form of protest and the position of higher-level governments are two crucial conditions that explain the difference in government strategies in the pattern 1 and pattern 2 cases.

The pair-wise comparison of the cases in pattern 2 and pattern 3

One crucial difference in the cases in pattern 2 and pattern 3 is that local governments in the former continued the debated projects, whereas local governments in the latter relocated them. In Table 7.8, it seems that no single condition was important in explaining this difference.

However, if these two patterns are compared in a more detailed way, it can be concluded that the stage of the projects and the position of higher-level governments are two crucial conditions that explain the difference in the cases between pattern 2 and pattern 3. As to the stage of projects, all the debated projects in the cases in pattern 3 were in their early stage. Four of them, namely, the Panyu, the Songjiang, the Tianjingwa, and the Liulitun waste incineration power plant were in their planning stage. The Xiamen PX plant was in its early construction stage. It was much less costly to relocate these projects compared to the three projects in the cases in pattern 2: the Dalian PX case, the Pengzhou refinery plant, and the Kunming refinery plant (see the section "Understanding the similarities in the three cases in pattern 2" above). When the costs of removing the debated projects were too high for local governments, they tended to stick to their initial strategies – mostly the continuation of the debated plants. If the costs of project relocation were not that high, they tended to relocate them.

The second condition is the position of higher-level governments. As shown above, the higher-level governments generally supported the three projects: the Kunming refinery plant, the Pengzhou refinery plant, and the Dalian PX plant. Local governments finally tended to continue them. Regarding the five projects in the cases in pattern 3, the higher-level governments did not publicly express their support. Then local governments tended to relocate the projects.

It can be concluded that the stage of projects and the position of higher-level governments are two important conditions in explaining the difference in the cases in pattern 2 and pattern 3.

Toward an explanation of the differences in the three patterns of government strategies during environmental conflicts

Three pair-wise comparisons were made in the above three subsections. In general, the three conditions, namely, the position of higher-level government, the form of protest, and the stage of the project, are important for explaining the differences in the three patterns of government strategies. Regarding the other four conditions, namely, the position of the national mass media, the scale of protest, the involvement of the activists, and the occurrence of events, it has been shown that they are not as important as the earlier three in explaining the differences in the three patterns of government strategies. The crucial conditions that differentiate the three patterns of government strategies are presented in Table 7.9.

Regarding the difference between project cancellation and project relocation, the form of protest really matters. As for the difference between project continuation and project cancellation, two conditions, namely, the form of protest and

Table 7.9 The conditions that explain the differences in the three patterns of government strategies

Crucial condition	Pattern of government strategies
The form of protest	Project cancellation/project relocation
The form of protest and the position of higher-level governments	Project continuation/project cancellation
The stage of the project and the position of higher-level governments	Project relocation/project continuation

the position of higher-level governments, matter. For the difference between project relocation and project cancellation, two conditions, namely, the stage of the project and the position of higher-level governments, matter.

Conclusions

In this chapter, an in-depth qualitative comparative study of 10 cases of environmental conflicts was reported (see also Li et al. 2017). Using both the method of agreement and the method of difference, two different types of comparisons were made: within-pattern comparison and cross-pattern comparison. Some most important conclusions are drawn as follows.

First, three patterns regarding government strategies during environmental conflicts were identified as leading to three different outcomes: project continuation, project abandonment, and project relocation.

Second, the three within-pattern comparisons using the method of agreement elucidated why different cases show a similar pattern of government strategies. Three explanations were identified: (1) the occurrence of strong violent protests, the absence of strong support from higher-level governments and the national mass media, the absence of events, and the absence of activists are important in explaining the first pattern of government strategies, which results in project abandonment; (2) the absence of strong violent protests, the support of higher-level governments, and late stage of the projects (or high costs of project abandonment or removal) are crucially important in explaining the second pattern of government strategies, which leads to project continuation; and (3) the early stage of the projects, the absence of support from higher-level governments, the absence of support from the national mass media, and the absence of strong violent protests are important in explaining the third pattern of government strategies, which results in project relocation.

Third, three cross-pattern comparisons using the method of difference were made, and the crucial differences between the patterns of government strategies were explained. Three findings were identified: (1) the form of protest is an important condition in explaining the difference between project abandonment and project relocation; (2) the stage of the project and the position of higher-level governments are important conditions in explaining the difference between project continuation and project relocation; and (3) the form of protest and the position of higher-level governments are important conditions in explaining the difference between project continuation and project abandonment.

In general, the comparative study in this chapter uses two basic comparative methods: the method of agreement and the method of difference. These two methods allowed me to look at the 10 cases from a higher aggregation level to seek an explanation for the application of government strategies – the patterns of government strategies. It has one added value compared to the two single case studies in Chapters 5 and 6: the relative importance of the conditions in explaining the application of government strategies can be identified.

In addition, the two methods are different. The method of agreement allows a focus on a specific pattern to seek an explanation about what matters for similarities among cases. The method of difference looks for an explanation for the differences in various patterns. In other words, the former reveals what conditions are important in explaining the same result: project cancellation, project relocation, and project continuation, whereas the latter reveals the conditions that are crucial in explaining differences in results. In the next chapter, I further study how the conditions, mostly the relatively important conditions identified in this chapter, work in a conjunctural way in explaining the application of government strategies.

Notes

1 http://news.ifeng.com/mainland/detail_2012_10/30/18656522_0.shtml, available on April 28, 2015.
2 http://news.ifeng.com/mainland/detail_2012_10/30/18656522_0.shtml, available on April 28, 2015.
3 www.boxun.com/news/gb/china/2011/05/201105301321.shtml, available on May 29, 2015.
4 www.boxun.com/news/gb/china/2011/05/201105301321.shtml, available on May 29, 2015.
5 www.epochtimes.com/b5/11/4/11/n3224806.htm, available on April 28, 2015.
6 www.boxun.com/news/gb/china/2011/05/201105301321.shtml, available on April 28, 2015.
7 http://news.ifeng.com/mainland/detail_2012_10/30/18656522_0.shtml, available on May 29, 2015.
8 www.greening-china.com/CaseCenter/show.php?itemid=166, available on April 28, 2015.
9 www.boxun.com/news/gb/china/2011/05/201105301321.shtml, available on May 6, 2015.
10 http://blog.sina.com.cn/s/blog_4b8bd1450102e9eo.html, available on May 29, 2015.
11 http://blog.sina.com.cn/s/blog_4b8bd1450102e9eo.html, available on May 29, 2015.
12 http://blog.sina.com.cn/s/blog_4b8bd1450102e9eo.html, available on May 29, 2015.
13 http://news.qq.com/a/20110531/000478.htm, available on May 29, 2015.
14 http://news.qq.com/a/20110531/000478.htm, available on May 29, 2015.
15 http://blog.sina.com.cn/s/blog_4b8bd1450102e9eo.html, available on May 29, 2015.
16 www.bbc.co.uk/zhongwen/simp/chinese_news/2011/08/110814_dalian_demo_closure.shtml, available on May 29, 2015.
17 www.bbc.co.uk/zhongwen/trad/china/2013/05/130516_kunming_refinery_protest_reflection, available on May 29, 2015.

18 www.bbc.co.uk/zhongwen/trad/china/2013/05/130504_china_kunming_protest_environment.shtml?MOB, available on May 29, 2015.
19 www.bbc.co.uk/zhongwen/simp/china/2013/05/130510_china_kunming_pxnews.shtml, available on May 29, 2015. www.bbc.co.uk/zhongwen/simp/chinese_news/2011/08/110814_dalian_demo_closure.shtml, available on May 29, 2015.
20 www.npr.org/blogs/thetwo-way/2013/05/04/181154978/to-silence-discontent-chinese-officials-alter-calendar, available on April 28, 2015.
21 www.npr.org/blogs/thetwo-way/2013/05/04/181154978/to-silence-discontent-chinese-officials-alter-calendar, available on April 28, 2015.
22 http://focus.cnhubei.com/columns/columns4/201109/t1830576.shtml, available on May 29, 2015.
23 http://news.sina.com.cn/c/2011-09-29/171223239636.shtml, available on May 29, 2015.
24 www.guancha.cn/Project/2013_05_12_144083.shtml, available on May 29, 2015.
25 http://magapp.caixin.com/2013-05-11/100526932.html, available on May 29, 2015.
26 http://magapp.caixin.com/2013-05-11/100526932.html, available on May 29, 2015.
27 www.infzm.com/content/92816, available on May 29, 2015. http://news.sina.com.cn/pl/2013-05-15/070927125012.shtml, available on May 29, 2015.
28 www.lifeweek.com.cn/2011/0826/34723_6.shtml, available on May 26, 2015.
29 http://ucwap.ifeng.com/news/dalu/news?vt=5&aid=70763930&mid=, available on May 29, 2015.
30 www.bbc.co.uk/zhongwen/simp/china/2013/03/130330_kunming_petro.shtml, available on May 29, 2015.
31 http://magapp.caixin.com/2013-05-11/100526932.html, available on May 29, 2015.
32 www.lifeweek.com.cn/2011/0826/34723_5.shtml, available on May 26, 2015.
33 www.guancha.cn/Project/2013_05_12_144083.shtml, available on May 29, 2015.
34 www.fortuneconferences.com/global-forum-chinese-2013/, available on May 29, 2015. http://roll.sohu.com/20130527/n377130732.shtml, available on May 29, 2015.
35 In this book, I use the stage of protest as a condition in explaining the application of government strategies during environmental conflicts. In fact, the analysis here means that the perceived costs of project cancellation might be a more useful indicator to strengthen my arguments. The problem is that the costs of project cancellation cannot be clearly calculated. Thus, I use the indicator of the stage of the project rather than the costs of project cancellation in this analysis.
36 http://blog.sina.com.cn/s/blog_4af9923001015j7r.html, available on May 29, 2015.
37 http://bbs.tianya.cn/post-free-1566273-1.shtml, available on May 29, 2015.
38 www.rfa.org/cantonese/news/incinerator-06142012102543.html, available on May 29, 2015.
39 www.nbweekly.com/news/observe/200904/9614.aspx, available on May 29, 2015.
40 http://gz.oeeee.com/a/20091204/812923.html, available on May 29, 2015. Local media reported that hundreds of local residents attended this protest. Here, the number of 500 was established to indicate the approximate scale of protest.
41 http://bbs.tianya.cn/post-free-1566273-1.shtml, available on May 29, 2015.
42 www.washingtonpost.com/wp-dyn/content/article/2007/06/27/AR2007062702962.html, available on May 29, 2015.
43 http://politics.people.com.cn/GB/14562/13874057.html, available on May 29, 2015.
44 http://business.sohu.com/20091203/n268644939.shtml, available on May 29, 2015.
45 http://energy.people.com.cn/GB/10504196.html, available on May 29, 2015.
46 www.infzm.com/content/9650, available on May 29, 2015.
47 http://news.cctv.com/society/20070416/102000.shtml, available on April 28, 2015.
48 http://news.ycwb.com/2009-11/22/content_2338256.htm, available on June 24, 2015.
49 http://news.sina.com.cn/c/2007-12-28/053914619388.shtml, available on May 29, 2015.
50 http://blog.sina.com.cn/s/blog_683c74c901016npo.html, available on May 29, 2015.

51 http://news.xinhuanet.com/politics/2011-04/15/c_121309085.htm, available on May 29, 2015.
52 http://news.xinhuanet.com/politics/2011-04/15/c_121309085.htm, available on May 29, 2015.
53 http://news.sina.com.cn/c/2007-12-28/053914619388.shtml, available on May 29, 2015.
54 http://news.sina.com.cn/c/2007-12-28/053914619388.shtml, available on May 29, 2015.
55 www.13332888731.com/news/46.html, available on May 29, 2015.

8 Under what conditions do Chinese local governments make compromises with local communities during environmental conflicts?

Introduction

This chapter aims to research the conditions under which Chinese local governments make compromises with local communities during environmental conflicts. Crisp-set qualitative comparative analysis (csQCA) is used as a method to compare 10 cases of environmental conflicts, allowing the exploration of how combinations of four conditions, namely, the scale of protest, the form of protest, the position of the central government, and the stage of projects, result in the occurrence and nonoccurrence of government compromises. The occurrence of government compromises implies the application of a giving in, a collaboration, or a facilitation strategy, whereas the nonoccurrence of government compromises implies the adoption of a go-alone, a suppression, or a tension reduction strategy. One added value of this chapter is that it elucidates how the conditions work in a conjunctural way to influence the application of government strategies during environmental conflicts. This chapter is structured in four sections.

First, the conceptual framework used in this chapter is elaborated. The calibration of the four conditions and the outcomes is presented in the next section followed by the analysis and results. Finally, conclusions and discussion are presented.

Conceptual framework

The following research question is answered in this chapter:

- *Under what conditions do Chinese local governments make compromises with local communities during environmental conflicts?*

To answer the above research question, qualitative comparative analysis (QCA) is applied. In Chapters 5 and 6, two single case studies were reported to elucidate why local governments adopted different strategies in environmental conflicts. In these two studies, the occurrence of various government strategies was explained. In-depth knowledge about the application of government strategies in environmental conflicts was obtained. One limitation of these two studies is that the conclusions drawn might not be generalizable. In Chapter 7, a comparative case study was reported. This facilitated identification of the conditions that were relatively important in explaining the similarities and differences in the patterns of government strategies during environmental conflicts. However, it did not reveal how various conditions explain the application of government strategies in a conjunctural way. QCA is a good option to research this. Another practical reason for using QCA is its appropriateness for medium-N studies (10–50 cases) (Vis 2012). Using QCA allows comparison of the 10 current cases in a systematic and structured way. One crucial added value of QCA compared to the comparative case study in Chapter 7 is that it facilitates studying how combinations of the identified conditions influence the application of government strategies in environmental conflicts.

In this chapter, csQCA is the method used; the reasons for its choice were introduced in Chapter 4. The adopted method having been established, the conceptual framework used is presented below; this framework is adapted from the conceptual framework constructed in Chapter 3.

The outcome to study: the occurrence/nonoccurrence of government compromises during environmental conflicts

In the conceptual framework constructed in Chapter 3, six government strategies were identified, namely, go-alone, suppression, tension reduction, giving in, collaboration, and facilitation. Using csQCA, however, requires the dichotomization of the outcomes of interest (Ragin 1987). The dichotomization process should be based on both adequate prior theoretical knowledge and empirical insights (Wagemann and Schneider 2010). These six government strategies can be generally dichotomized into two outcomes: the nonoccurrence of compromise and the occurrence of compromise. In this chapter, compromise is defined as *the change in decisions made by local governments regarding the substantive outcomes of the debated projects.* The first three government strategies (go-alone, suppression, and tension reduction) imply that local governments generally dominate the decision-making processes without altering their original decisions substantially. It should be noted that the application of a tension reduction strategy implies that local governments may make compromises. The compromises are temporary and unimportant, however. The last three government strategies (giving in, collaboration, and facilitation) imply that compromises are made by local governments, and government decisions are changed. Specifically, the application of a giving in strategy means that local governments make big compromises and they may give up the debated projects. The application of a collaboration or a facilitation strategy means that local governments seek a win–win solution to solve a problem, implying that they make compromises that combine their own objectives with those of others.

The conditions that explain the occurrence/nonoccurrence of compromises by local governments

For csQCA, there are 2^n combinations for N conditions (Rihoux and Ragin 2009). Because of the limited diversity problem, the observed cases may occupy only a small portion of the potential combinations of conditions (Ragin 1987). For example, if I have 10 cases and choose five conditions, there are 2^5 (32) potential combinations of conditions. It is therefore better to choose a limited or moderate number of conditions based on the balance between the number of conditions and the number of cases (Rihoux and Ragin 2009).

In Chapter 3, seven conditions were identified as influencing the application of government strategies, namely, the form and scale of protest, the position of higher-level governments, the position of the national mass media, the involvement of activists, the occurrence of events, and the stage of projects. From the analysis in Chapter 7, compared to the other conditions, three conditions, namely, the position of higher-level governments, the stage of projects, and the form of protest, are relatively important in explaining differences in the patterns of government strategies in environmental conflicts.

In this chapter, I merge the position of higher-level governments and the position of the national mass media into a higher-order construct: the position of central government (see Rihoux and Ragin 2009). There are two reasons for doing this. First, several positions may be taken by higher-level governments, including both provincial governments and central governments. It is necessary to focus on one level of government. Normally, the position of central government is more crucial than that of provincial governments in influencing strategies applied by local governments. Thus, I focus on the position of the national government. Second, the national mass media in China are affiliated to the central government, and they can be viewed as government agencies. Furthermore, after iterative analysis of csQCA, I found that it was possible to use a fourth condition, the scale of protest, to explain the occurrence and nonoccurrence of government compromises in environmental conflicts. The literature shows that the scale of protest influences the application of government strategies. Protests involving a large number of participants imply "big trouble" for local governments (Cai 2002), as they may disrupt social disorder (Cai 2004). As a result, local governments tend to adjust their strategies to the demands of local citizens.

To conclude, four conditions are chosen in this chapter to explain the occurrence and nonoccurrence of government compromises: the scale of protest, the form of protest, the position of the central government, and the stage of projects. The raw data are presented in Table 8.1.

Calibration of outcome and conditions

In this section, I calibrate the outcome and conditions studied. Calibration in csQCA is about defining the qualitative difference in kind that characterizes the underlying concepts, resulting in the dichotomization of conditions and outcomes

Table 8.1 Raw data matrix of the 10 cases of environmental conflicts

Case	Stage of the project	Form of protest	Scale of protest	Position of the central government	Outcome
NB	Planning stage	Strong violence	1,000	Silence	Project cancellation
XM	Initial construction stage	Peaceful	8,000–10,000	Opposition	Project cancellation
DL	Operation stage	Peaceful	12,000	Support	Project continuation
KM	Planning stage	Peaceful	2,000	Support	Project continuation
PZ	Before operation	Peaceful	0	Support	Project continuation
PY	Planning stage	Peaceful	500	Opposition	Project relocation
LLT	Planning stage	Peaceful	1,000	Opposition	Project relocation
TJW	Planning stage	Strong violence	100	Contradictory	Project relocation
SJ	Planning stage	Peaceful	600	Silence	Project relocation
WX	Trial operation	Strong violence	10,000	Silence	Project cancellation

Source: Li et al. (2016b).

Notes
NB = Ningbo case, XM = Xiamen case, DL = Dalian case, KM = Kunming case, PZ = Pengzhou case, PY = Panyu case, LLT = Liulitun case, TJW = Tianjingwa case, SJ = Sonjiang case, and WX = Wuxi case.

(Ragin 1987). How set membership scores are assigned is crucial for QCA, especially the specification of the qualitative anchors (Schneider and Wagemann 2010). The codings "1" and "0" in csQCA connote the meaning of "fully-in" and "fully-out," respectively; this means that cases with a set membership score of 1 and 0 are different in kind (Wagemann and Schneider 2010). In the following, the outcome studied is first calibrated, and then the four conditions are described.

Outcome: the occurrence and nonoccurrence of compromises by Chinese local governments with local communities during environmental conflicts

The occurrence of compromise means that local governments change their original decisions, whereas the absence of compromise implies that they stick to their original decisions. The substantive outcome of the debated projects is the indicator chosen to calibrate the outcome: the occurrence and nonoccurrence of government compromises. Three substantive outcomes are identified: project relocation, project

Table 8.2 The membership scores with regard to the occurrence of compromises by Chinese local governments in 10 cases

Set	Indicator	Case	Score
The occurrence of compromises	Project relocation	LLT, TJW, SJ, PY	1
	Project cancellation	WX, NB, XM	
The nonoccurrence of compromises	Project continuation	DL, PZ, KM	0

cancellation, and project continuation. The cases calibrated as 1 are indicated by project relocation *or* project cancellation. The cases calibrated as 0 are indicated by project continuation. The membership scores of the 10 cases regarding the occurrence of government compromises are presented in Table 8.2.

Condition 1: the scale of protest

The number of participants in protests is the chosen indicator to calibrate the condition: the scale of protest. In some cases, more than one protest occurred. The number of participants in protests is mainly derived from the maximum number of participants involved in protests as reported in mass media inside and outside China.

The number of 5,000 is chosen as the cross-over point to dichotomize case membership scores in the set: the presence of large-scale protests versus the absence of large-scale protests. This cross-over point depends on the data distribution regarding the number of participants (see Table 8.1): I observed a big gap in the cases between the numbers 2,000 and 8,000. Second, cluster analysis using the QCA software TOSMANA (Cronqvist 2017) identified 5,000 as the cross-over point. For the Xiamen case, the reported numbers of participants range from 8,000 to 10,000. The figure of 8,000 is used as the number of participants in the Xiamen protest. The case scores are robust: the value of 10,000 participants indicates the number of 6,000 as the cross-over point, but this does not influence the calibration. The membership scores in terms of the scale of protest in the 10 cases are presented in Table 8.3.

Table 8.3 The membership scores with regard to the scale of protest in the 10 cases

Set of protest	Indicator	Case	Score
The presence of large-scale protests	Maximum number of participants over 5,000	XM, DL, WX	1
The absence of large-scale protests	Maximum number of participants less than 5,000	NB, PY, LLT, SJ, PZ, TJW, KM	0

Notes
NB = Ningbo case, XM = Xiamen case, DL = Dalian case, KM = Kunming case, PZ = Pengzhou case, PY = Panyu case, LLT = Liulitun case, TJW = Tianjingwa case, SJ = Sonjiang case, and WX = Wuxi case.

Condition 2: the form of protest

The set is calibrated into the presence of violent protest and the absence of violent protest. The occurrence of casualties or injuries indicates violent protest (calibrated as 1) and their nonoccurrence indicates nonviolent protest (calibrated as 0). The membership scores for the 10 cases in terms of the form of protest are presented in Table 8.4.

It should be noted that a strong violent protest occurred in the Tianjingwa case, in which some persons from the Environmental Protection Department of the Jiangsu provincial government used violent force to suppress local residents. However, the influence of this violent protest on the application of local governments' strategies in Nanjing was limited (see Chapter 7, "Within-pattern comparison using the method of agreement"). Consequently, the Tianjingwa case was calibrated as 0.

Condition 3: the stage of projects

Four project stages are identified: the planning stage, the early construction stage, the final construction stage, and the formal operation stage. The projects in the first two stages are not different in kind (*quality*) as their investments are mostly limited. In this chapter, these two stages are termed as the early stage. The projects in the latter two stages are not different in kind either because large investments have been made, making compromises difficult. These two stages are defined as the late stage. The cases qualified as being in the early stage are calibrated as 1, and the cases qualified as being in the late stage are calibrated as 0. The membership scores of the 10 cases regarding the stage of projects are shown in Table 8.5.

The Kunming case is striking. Although the PX plant in this case was in its planning stage, local governments in Kunming had made large investments in it. A relocation or a giving in strategy would have been very costly for these governments. Consequently, it is argued that, although this PX plant was in its planning stage, the costs of changing the original plans for the project were high. The condition in the Kunming case was therefore calibrated as 0.

Table 8.4 The membership scores with regard to the form of protest in the 10 cases

Set	Indicator	Case	Score
The presence of violent protest	Strong violence	NB, WX	1
The absence of violent protest	No/little/some violence	DL, PZ, LLT, PY, XM, KM, SJ, TJW	0

Notes
NB = Ningbo case, XM = Xiamen case, DL = Dalian case, KM = Kunming case, PZ = Pengzhou case, PY = Panyu case, LLT = Liulitun case, TJW = Tianjingwa case, SJ = Sonjiang case, and WX = Wuxi case.

Table 8.5 The membership scores with regard to the stage of projects in the 10 cases

Set	Indicator	Case	Score
Early stage	Planning	NB, PY, LLT, TJW, SJ	1
	Early construction	XM	
Late stage	Before operation/trial operation/ high costs of preparatory work	PZ, WX, KM	0
	Formal operation	DL	

Notes
NB = Ningbo case, XM = Xiamen case, DL = Dalian case, KM = Kunming case, PZ = Pengzhou case, PY = Panyu case, LLT = Liulitun case, TJW = Tianjingwa case, SJ = Sonjiang case, and WX = Wuxi case.

Condition 4: the position of the Chinese central government

Four positions held by the Chinese central government on debated projects during environmental conflicts are identified: support, silence, contradiction, and opposition. *Support* means that the Chinese central government legitimized the advancement of the debated projects. *Silence* implies that it did not publicly show its position. *Contradiction* implies that different government agencies from the central government showed various, mostly contradictory positions. *Opposition* means that it advised *or* ordered local governments to reconsider the debated projects. Cases regarding the position of the Chinese central government are calibrated into the presence of support and the absence of support. One issue that should be mentioned is that the Chinese central government may hold different positions in one case. In this chapter, I chose only its position after the occurrence of protests because it is the Chinese central government's post-protest attitude that often shows its real intentions, and this is the important reference toward which local governments to orient themselves.

The positions of the State Council's ministries, or of the Chinese national mass media (most notably the *People's Daily*, China Central Television (CCTV), and the Xinhua News Agency), are used to calibrate the condition. A case calibrated as 1 signals that the ministries or the national mass media publicly supported the advancement of the projects; a case calibrated as 0 denotes the absence of support. The membership scores regarding the position of the Chinese central government after the occurrence of protests in the 10 cases are presented in Table 8.6.

The above five subsections elaborately show the calibration processes of outcome and conditions studied in this chapter. Based on them, the data matrix of the 10 cases is presented as Table 8.7.

Analysis

A truth table, a presentation of logically possible combinations (or configurations) of conditions, is used to display the observed combinations of conditions

Table 8.6 The membership scores with regard to the position of the Chinese central
government in the 10 cases

Set	Indicator	Case	Score
The presence of support	Support	DL, PZ, KM	0
The absence of support	Contradiction	TJW	1
	Silence	NB, WX, SJ	
	Opposition	XM, LLT, PY	

Notes
NB = Ningbo case, XM = Xiamen case, DL = Dalian case, KM = Kunming case, PZ = Pengzhou
case, PY = Panyu case, LLT = Liulitun case, TJW = Tianjingwa case, SJ = Sonjiang case, and WX =
Wuxi case.

Table 8.7 Data matrix

Row	Case	Conditions				Outcome
		L	V	E	S	C
1	NB	0	1	1	0	1
2	XM	1	0	1	0	1
3	DL	1	0	0	1	0
4	KM	0	0	0	1	0
5	PZ	0	0	0	1	0
6	PY	0	0	1	0	1
7	LLT	0	0	1	0	1
8	TJW	0	0	1	0	1
9	SJ	0	0	1	0	1
10	WX	1	1	0	0	1

Source: Li et al. (2016b).

Notes
L = large-scale protest, V = violent protest, E = early stage of the project, S = support from the
Chinese central government, C = occurrence of compromises by local governments.

(Ragin 1987). Four conditions are chosen, implying that there are 16 (2^4) possible combinations of conditions. There are six rows in Table 8.8 and 10 logical
remainders. Logical remainders are rows in a truth table without empirical
information, implying that the combinations of conditions do not exist empirically (Ragin 2000).

The main interest is to explain both the *occurrence* and the *nonoccurrence*
of government compromises in environmental conflicts. Asymmetric causality
means that the explanation of the former does not say much about the latter
(Schneider and Wagemann 2012). It is necessary therefore to analyze the
occurrence and the nonoccurrence of compromises by local governments
separately.

Table 8.8 Truth table

Row	Conditions					Outcome	Case
	L	*V*	*E*	*S*	*C*		
1	0	1	1	0	1		NB
2	1	0	1	0	1		XM
3	0	0	1	0	1		PY, LLT, TJW, SJ
4	1	1	0	0	1		WX
5	1	0	0	1	0		DL
6	0	0	0	1	0		PZ, KM

Source: Li et al. (2016b).

Notes
L = large-scale protest, V = violent protest, E = early stage of the project, S = support from the Chinese central government, C = occurrence of compromises by local governments.
NB = Ningbo case, XM = Xiamen case, DL = Dalian case, KM = Kunming case, PZ = Pengzhou case, PY = Panyu case, LLT = Liulitun case, TJW = Tianjingwa case, SJ = Sonjiang case, and WX = Wuxi case.

Explaining the occurrence of compromise by local governments with local communities during environmental conflicts

Analysis of the truth table is primarily geared toward uncovering sufficient conditions. Before analyzing the sufficiency of conditions, I first analyze the consistency and coverage of necessary conditions. The formal expression of a necessary condition hypothesis is "Y only if X" (Dul et al. 2010). A condition X is necessary if, whenever the outcome Y is present, the condition is also present (Schneider and Wagemann 2012, 69). The consistency of a necessary condition assesses the degree to which this condition overlaps with a particular outcome relative to all cases with the same outcome (Ragin 2000). The coverage for necessary conditions is mostly about the relevance of the necessary conditions. High coverage of a necessary condition implies that it is a relevant necessary condition. I establish a consistency threshold of 1, which is the mostly widely criterion for csQCA analysis (Schneider and Wagemann 2012).

The results of the analysis in Table 8.9 show that the absence of support from the central government has a consistency of 1, implying that it is a necessary condition in explaining the occurrence of government compromises during environmental conflicts. This implies that, when local governments make compromises with local communities, they never receive support from the central government. In other words, it can be said that the absence of support from central government is a prerequisite condition that explains the absence of government compromises with local communities in environmental conflicts. The other conditions have a low consistency ranging from 0.14 to 0.86.

After analyzing necessary conditions, I analyze sufficient conditions. The formal expression of sufficient condition is "if X, then Y," which means that, if X occurs, Y will follow (Ragin 2000). Because of limited diversity, the analysis

Table 8.9 Analysis of necessary conditions for the occurrence of government compromises during environmental conflicts

Conditions	Consistency	Coverage
LARGE	0.29	0.67
large	0.71	0.71
EARLY	0.86	1.00
early	0.14	0.25
SUPPORT	0.00	0.00
support	1.00	1.00
VIOLENT	0.29	1.00
violence	0.71	0.63

Source: Li et al. (2016b).

based on a same truth table can yield different solution formulas (the results of truth table analysis QCA) depending on the assumptions about logical remainders (Ragin 2000).

In general, three solutions can be produced: the conservative solution, the most parsimonious solution, and the intermediate solution (Rihoux 2006; Wagemann and Schneider 2010). The conservative solution implies that the analysis does not rely on any assumptions about logical remainders and all of them are set as negative outcomes. The most parsimonious solution means that it is allowed to engage in thought experiments to take into account all simplifying assumptions[1] in order to achieve the most parsimonious solution (Ragin 2000). The intermediate solution only incorporates easy counterfactuals.[2] In addition, good QCA practice demands that all three solution formulas be shown (Rihoux 2006). Often, the conservative solution tends to be too complex to be plausibly interpreted, and the most parsimonious solution may be unrealistically simple or contradictory to our directional expectations, which are theory-guided hunches about the relationships between the conditions and the outcomes of interest. The intermediate solution is criticized as well because it makes the distinction between theory and the analysis unclear (Baumgartner 2014). The fsQCA software[3] is used to analyze the truth table (see Table 8.8). In the following, the three solution formulas as well as the assumptions about logical remainders are presented; this allows readers to make their own judgments about the plausibility of each solution formula (Schneider and Wagemann 2012). The analysis first produces the conservative solution formula as follows:

$$s \times E \times 1 + v \times E \times s + V \times e \times s \times L \rightarrow C$$

Meanwhile, the most parsimonious solution formula is produced through the process of Boolean minimization[4] and all simplifying assumptions are used (Ragin 2008).

$$s \rightarrow C$$

The conservative and most parsimonious solution formulas having been produced, directional expectations are used to craft the intermediate solution formula. In Chapter 3, it was proposed that local governments tend to apply a tension reduction or a giving in strategy to cope with large-scale protests and violent protests (proposition 1 and proposition 2 in the section "Propositions regarding the application of government strategies during environmental conflicts"). In addition, it was proposed that local governments tend to apply a tension reduction, a giving in, a collaboration, or a facilitation strategy when the debated projects are in the early stage and when higher-level governments criticize the continuation of the debated projects (proposition 4 and proposition 5). These propositions show that the directional expectations could equally plausibly go into both directions. No meaningful directional expectations can be formulated. Consequently, it is not possible to develop the intermediate solution formula. The conservative solution formula is interpreted in the following as it is not too complex to be interpreted.

The conservative solution formula includes three paths, and their raw coverage and unique coverage are presented in Table 8.10. Raw coverage indicates the proportion of the outcome that is covered in a single path (Ragin 2008). Unique coverage indicates the proportion of the outcome that is uniquely covered by a single path (Schneider and Wagemann 2012). The three paths have the same unique coverage 0.14, but not the same raw coverage. In the following, the three paths and their corresponding cases are elaborated in detail.

Path 1: the combination of the presence of large-scale (L) and strong violent protests (V), late stage of the project (e), and absence of support from the Chinese central government (s) leads to the occurrence of compromises by local governments during environmental conflicts

This path identified three INUS conditions; these are *insufficient* but *non-redundant* parts of an *unnecessary* but *sufficient* condition (see Mackie 1980),

Table 8.10 Conservative solution formula for the occurrence of government compromises during environmental conflicts

Path	$V \times e \times s \times L$	$v \times E \times s$	$s \times E \times l$
Case	WX	XM, SJ, TJW, PY, LLT	PY, LLT, NB, SJ, TJW
Raw coverage	0.14	0.71	0.71
Unique coverage	0.14	0.14	0.14
Solution coverage	1		
Solution consistency	1		

Source: Li et al. (2016b).

Notes
NB = Ningbo case, XM = Xiamen case, DL = Dalian case, KM = Kunming case, PZ = Pengzhou case, PY = Panyu case, LLT = Liulitun case, TJW = Tianjingwa case, SJ = Sonjiang case, and WX = Wuxi case.

namely, the presence of violent protests (V), large-scale protests (L), and late stage of the project (e). This path covers the Wuxi case. The waste incineration power plant had finished its trial operation. Abandoning it would be very costly for local governments. It was reported that over 4,000 anti-riot police were dispatched by local governments in Wuxi to suppress local citizens. The Chinese state is sensitive to the occurrence of large-scale violent confrontations. If they occurred, the tension between local residents and local governments would be very high. In the Wuxi case, after the large-scale violent confrontation, relevant information about it was censored by the Chinese state. Large-scale violent confrontation between local governments and citizens is a political taboo, after which the state normally tends to keep silent, and discussion about the Wuxi case was inhibited. If the waste incineration power plant was further advanced, then the large-scale violent confrontation in Wuxi might be widely discussed around China, and this would significantly damage state legitimacy. To avoid this, local governments in Wuxi would prefer to compromise, even though this was costly.

Path 2: the combination of early stage of the project (E), absence of violence (v), and absence of support from the central government (s) results in the occurrence of government compromises during environmental conflicts

This path identified two INUS conditions: early stage of the project (E) and absence of violent protest (v). Five cases are covered by the second path: the Panyu, the Songjiang, the Xiamen, the Tianjingwa, and the Liulitun case. In these cases, no strong violent protests occurred. In addition, the central government did not publicly show its support for these debated projects in the five cases after protests. In the Songjiang case, the central government kept silent after the occurrence of protest. In the Panyu case, a national mass medium, CCTV, publicly reported the debate concerning the construction of the Panyu waste incineration power plant and commented that local governments in Guangzhou should enhance public participation. In the Xiamen case, the State Environmental Protection Administration (SEPA) advised Xiamen Municipality to make a comprehensive environmental impact assessment for the whole Haicang region. In the Liulitun case, SEPA demanded that Beijing Municipality postpone the Liulitun waste incinerator. In the Tianjingwa case, both the Ministry of Environmental Protection (MEP)[5] and the Ministry of Housing and Urban-Rural Development (MHURD) were involved. The former supported the continuation of the Tianjingwa waste incineration power plant, whereas the latter investigated the project but did not show its position. In short, no violent protests occurred in the five cases, and the absence of support from the central government in the five cases gave local governments little confidence to further advance the debated projects. In addition, all the debated projects in the five cases were in their early stage, implying that abandoning or relocating them was not that costly for local governments. As a result, they relocated or cancelled them.

Path 3: the combination of absence of support from the central government (s), early stage of the project (E), and absence of large-scale protests (l) results in the occurrence of government compromises during environmental conflicts

This path identified one INUS condition: the absence of large-scale protests (l). This path covers five cases: the Panyu, the Tianjingwa, the Ningbo, the Songjiang, and the Liulitun case. All five projects were in their early stage, and the national government did not show its support for their continuation. Furthermore, no large-scale protests occurred in these cases. About 600 local citizens expressed their opposition to the construction of the waste incineration power plant in Songjiang. In the Liulitun case, about 1,000 participants went to SEPA to express their opposition to the construction of the Liulitun waste incineration power plant. In the Panyu case, hundreds of local residents went to Guangzhou Municipality to show their opposition to the construction of the Panyu waste incineration power plant. In the Tianjingwa case, about 100 local residents assembled in front of the Jiangsu Provincial Environmental Protection Department, demanding a conversation with its key leaders. In the Ningbo case, about 1,000 local residents participated in the protest. In sum, the combination of small-scale protests, the early stage of the projects, and the absence of support from the central government results in the occurrence of government compromises with local communities during environmental conflicts.

The explanation of the absence of compromises by local governments with local communities during environmental conflicts

In terms of explaining the nonoccurrence of government compromise, I follow the same procedures of analysis as above. The consistency and coverage of the necessary conditions were first tested, and the result is presented in Table 8.11.

It shows that three conditions, presence of support from the national government (S), late stage of the project (e), and absence of violent protests (v), have a consistency of 1.00. This result can also be identified in the truth table in

Table 8.11 Analysis of necessary conditions for the absence of government compromises during environmental conflicts

Condition	Consistency	Coverage
LARGE	0.33	0.33
large	0.67	0.29
EARLY	0.00	0.00
early	1.00	0.75
SUPPORT	1.00	1.00
support	0.00	0.00
VIOLENT	0.00	0.00
violent	1.00	0.38

Source: Li et al. (2016b).

Table 8.8. It can be further understood as that, when local governments do not make compromises, there is always support from the central government at the late stage of projects after nonviolent protests. In addition, the necessity coverage of these three conditions is 1.00, 0.75, and 0.38; this means that the absence of government compromises occurs in all the cases that have the support of the Chinese central government after protests, in 75 percent of the cases where the debated projects are in the late stage, and in 38 percent of cases without violent protests. Given the low coverage of the condition, the absence of violent protests, it is not established as a relevant necessary condition: it might be a trivial necessary condition (Ragin 2008). Consequently, two relevant necessary conditions that explain the absence of government compromises during environmental conflicts can be identified: support from the central government (S) and late stage of the project (e).

The consistency and coverage of necessary conditions having been tested, the sufficient conditions that explain the absence of compromises by local governments are sought. Following the same procedure as above, the conservative solution formula is first produced as follows.

$$v \times S \times e \rightarrow c$$

It should be noted that there are two prime implicants, the end product of the logical minimization process: (1) support from the central government (S) and (2) late stage (e), and absence of violent protests (v) (v × e). The former is chosen given the high consistency and coverage of the condition, support of the central government (see Table 8.11). In fact, the choice of the two different prime implicants will result in the same conservative and intermediate solution formulas. The most parsimonious solution formula is produced in the following:

$$S \rightarrow c$$

To produce the intermediate solution formula, the directional expectations are that late stage of the project (e), absence of violent protests (v), absence of large-scale protests (l), and support of the Chinese central government (S) contribute to the absence of government compromises. In Chapter 3, I have proposed that local governments tend to apply a go-alone or a suppression strategy when higher-level governments (the Chinese central government in this chapter) support debated projects (see proposition 5 in the section "Propositions regarding the application of government strategies during environmental conflicts"). In addition, local governments tend to adopt a go-alone or a suppression strategy to cope with projects in their late stage (see proposition 4). And local governments tend to adopt a go-alone or a suppression strategy to deal with small-scale (here the absence of large-scale protest) and nonviolent protests (proposition 1 and proposition 2). As a result, the intermediate solution formula is developed as follows, based on the directional expectations:

$$v \times S \times e \rightarrow c$$

It can be seen that the conservative solution formula is the same as the intermediate solution formula. The conservative solution formula (also the intermediate solution formula) can be expressed as follows:

> The combination of support from the Chinese central government (S), absence of violent protests (v) and late stage of the project (e) leads to the absence of compromises by local governments during environmental conflicts.

In three cases – Dalian, Pengzhou, and Kunming – the Chinese central government expressed its support for the debated projects. After a large-scale protest, the National Development and Reform Commission (NDRC), jointly with four other ministries, released a notice that framed the occurrence of the large-scale protest in Dalian as a result of unsafe PX production. This implies that the central government did not deny the legitimacy of the continuation of the Dalian PX plant. In the Pengzhou and Kunming cases, two national mass media, *People's Daily* and CCTV, propagated the harmlessness of the PX project and the necessity of the refinery projects; this implies that the central government supported the continuation of the projects in both cases. Both the Dalian PX plant and the Pengzhou refinery plant were in their late stage, implying that it was costly for local governments to abandon them. Although the PX plant in the Kunming case was in its planning stage, some preparatory work had been done. An oil pipeline from Myanmar had been constructed, and this would have made compromises by local governments in Kunming very costly. Finally, no violent protests occurred in the three cases. This may have resulted in less pressure on local governments to make compromises. Thus, the combination of these three conditions results in the nonoccurrence of government compromises with local communities in the three cases. This finding partially corresponds with Cai's (2010) conclusion that, when citizens' demands are costly for local governments to resolve or protest actions initiated by citizens are not forceful, citizens have little possibility of succeeding in changing government decisions.

In this section, both the occurrence and the nonoccurrence of government compromises with local communities were analyzed using csQCA. Four causal recipes were identified, and this increased our knowledge about the explanation of why particular government strategies are applied during environmental conflicts.

Discussion and conclusion

In this chapter, I have attempted to answer the research question: under what conditions do Chinese local governments make compromises with local communities during environmental conflicts? I used csQCA as a method to address this question and some key conclusions are drawn.

The first conclusion is that the conclusions drawn in Chapter 7 are further specified through configurational thinking using csQCA. One finding identified in Chapter 7 was that the position of higher-level governments is important in explaining differences in patterns of government strategies. Specifically, the support of higher-level governments contributes to project continuation, whereas the absence of their support contributes to project relocation and project cancellation. This finding was specified in this chapter: the absence of support from the central government is a necessary condition that explains the occurrence of government compromises, and its presence is a necessary condition that explains their absence. This implies that, when local governments make compromises with local communities, they never receive support from the central government. However, if they stick to their original strategies, they always receive support from the central government.

In addition, it was found in Chapter 7 that the stage of the project is an important condition in explaining the differences in the patterns of government strategies: early stage of the project contributes to project relocation, whereas late stage contributes to project continuation. This chapter further specified this conclusion: the stage of the project (both early stage and late stage) is an INUS condition in explaining the occurrence of government compromises, and the late stage of the project is a relevant necessary condition for the absence of government compromises. This implies that the early or the late stage of the project, together with the other conditions, can lead to the occurrence of government compromises. In addition, the late stage of project is a necessary condition for the absence of compromise. This means that if there is an absence of government compromises during environmental conflicts, the debated projects are always in their late stage (or have high perceived costs of relocation or cancellation). In Chapter 7, it was found that the form of protest is an important condition in explaining the application of government strategies: strong violent protests contribute to project cancellation, whereas peaceful protests contribute to project relocation and project continuation. The analysis in Chapter 8 further specified this conclusion: the form of protest (both violent and nonviolent) is an INUS condition for the occurrence of government compromises, whereas the absence of violent protest is an irrelevant (or trivial) necessary condition for the nonoccurrence of government compromises. This implies that the presence or the absence of violent protests, combined with the other conditions, can lead to the occurrence of government compromises. When local governments stick to their original objectives, there are never violent protests. Finally, it was found in Chapter 7 that the scale of protest is not an important condition in explaining differences in patterns of government strategies during environmental conflicts. In this chapter, it was illustrated that the scale of protest is an INUS condition for the occurrence of government compromises. This implies that large-scale or small-scale protests, combined with the other conditions, can lead to the occurrence of government compromises.

The second conclusion is that the occurrence and the nonoccurrence of government compromise should be studied as two independent outcomes. Regarding the occurrence of government compromise, it seems that the Chinese state (including both central government and local governments) is responsive to citizens' concerns regarding the potentially negative influence of debated industrial projects on their health. Local governments sometimes even sacrificed economic interests for social stability (in the Wuxi case), and the central government even publicly rejected the decisions made by local governments when local citizens were dissatisfied with them (in the Liulitun case). Regarding the absence of government compromise, few positive signals regarding the application of government strategies have been seen: mostly the Chinese state had a very high intention to achieve the debated industrial projects despite strong opposition from the other actors.

In general, this chapter is a pilot attempt using csQCA to explain the application of government strategies in Chinese environmental conflicts (Li et al. 2016b). As in many csQCA analyses, the dichotomization of conditions and outcomes probably results in missing important information. Although some scholars argue that the elegance of simplicity is a big advantage of csQCA (Marx, Rihoux, and Ragin 2014), I experienced difficulty in dichotomizing conditions, making their calibration far from easy. For example, I had to calibrate the position of the central government during environmental conflicts. However, different ministries may hold various positions in a specific case, making it difficult to calibrate this condition. In the future, it would be worthwhile to use another variant of QCA, such as mvQCA or fsQCA, to study the application of government strategies on the basis of more fine-grained data and information. Another potential limitation of this study is that the dynamic dimensions of the 10 cases of environmental conflicts failed to be well explored (see Boswell and Brown 1999; Schneider and Rohlfing 2013). Government strategies in the 10 cases of environmental conflicts changed over time. More dynamic, in-depth comparative studies are needed to study this.

Notes

1 Simplifying assumptions are assumptions about logical remainders; they are used mainly to yield a solution formula that is less complex than the conservative solution formula (Schneider and Wagemann 2012).
2 Easy counterfactuals are defined as simplifying assumptions that are in line with both empirical evidence and directional expectations (Schneider and Wagemann 2012). In the process of producing the most parsimonious solutions, both the easy and the difficult counterfactuals are taken into account in order to obtain a simpler solution.
3 See: www.u.arizona.edu/~cragin/fsQCA/, Ragin, C., and Sean, D. (2014). fs/QCA [Computer Programme], Version 2.5. Irvine, CA: University of California.
4 Boolean minimization is the reduction of a long, complex expression into a shorter, more parsimonious one. Its basic expression is: If two Boolean expressions differ in only one causal condition yet produce the same outcome, then the causal condition that distinguishes the two expressions can be considered irrelevant and can be removed to create a simpler, combined expression (Ragin 1987, 93).
5 After 2008, SEPA was upgraded to MEP.

9 Toward a conclusion

Governing environmental conflicts in China

Introduction

It is now time to draw some final conclusions from the previous four empirical chapters. This chapter has three main aims. The first aim is to provide explicit answers to the research questions raised in Chapter 1. This will reveal what we have learned about the governance of environmental conflicts in China (see the section "Conclusions about research questions"). The second aim of this chapter is to summarize the contributions of this book, as well as to present reflections (see the section "Contributions and reflections"). The third aim is to elaborate what can be done in the next stage. To this end, a research agenda is set in the final section.

Conclusions about research questions

At the end of Chapter 1, the following three research questions were raised:

1   *What are the general characteristics of environmental conflicts in China? (What are environmental conflicts? Who are involved in environmental conflicts? What are their strategies? How do environmental conflicts evolve? And what are the outcomes of environmental conflicts?)*
2   *Which kinds of government strategies are applied by Chinese local governments in environmental conflicts?*
3   *How can the application of government strategies during environmental conflicts be explained?*

In the next subsections, the research questions are respectively answered.

What are the general characteristics of environmental conflicts in China?

Regarding the nature of environmental conflicts in China, this book provides the following insights:

1   *Various actors, such as local governments, central government, environmental NGOs, experts, local residents and activists, and private enterprises*

or state-owned enterprises (SOEs), are involved in Chinese environmental conflicts. Local governments are the first main category of actors to whom central government grants substantial discretion in governing local affairs, including environmental conflicts. Normally, various government agencies are involved in the governance of environmental conflicts, such as planning bureaus, city landscape bureaus, and development and reform bureaus. Among them, two agencies are highly relevant: the development and reform commission and the environmental protection bureau. The former is relatively more powerful than other bureaus and is mainly responsible for economic development, and the latter is in charge of the environmental impact assessment (EIA) of industrial projects. The Chinese central government is the second main actor. When local governments fail to address environmental conflicts, it will intervene. Environmental NGOs are the third category of actors involved. They usually do not publicly threaten state authorities or challenge government decisions. Often, they use their professional skills or knowledge to help the other actors to remedy perceived government mismanagement. Experts are the fourth category of actors. Sometimes, they are invited by local governments to persuade local residents to accept the decisions made by local governments (the Wuxi case). Sometimes, they publicly show their opposition to government decisions on the construction of industrial facilities (the Xiamen case). Activists and local residents are the fifth category of actors involved. Activists often coordinate the mobilized protests or design strategies for local residents to express their disapproval of government decisions. Private companies or SOEs are the sixth category of actors involved. They have a strong intention to achieve the construction of the industrial facilities to pursue economic interests. These six types of actors are the main actors involved in Chinese environmental conflicts. This is one important characteristic of environmental conflicts.

2   *Environmental conflict in China is a wicked problem.* The wickedness of environmental conflicts can be illustrated by the fact that the various actors involved have different perceptions regarding their nature and solution. Take waste incineration as an example. One government official (respondent 13) regarded conflicts concerning the construction of waste incineration power plants as a problem originating from the selfishness of citizens. She was convinced that waste incineration power plants must be built, and that removing the adjacent citizens was the best resolution. Another government official (respondent 3) argued that environmental conflict regarding the construction of waste incineration power plants is essentially a technological problem, because citizens worry about the potentially negative influence of waste incineration on their health. He argued that the technique of waste incineration was sophisticated and that waste incineration power plants should be advanced despite opposition from others. One respondent from a Beijing-based environmental NGO (respondent 8) claimed that the occurrence of environmental conflicts originates from the governance style of the

Chinese state: the Chinese state dominates the decision-making processes. Consequently, improvement of the decision-making style would be key in resolving environmental conflicts. My interviews with other respondents show that many of them have different or even contradictory views about the nature and resolution of conflicts regarding the construction of waste incineration power plants. Environmental conflict is therefore a wicked issue because of the lack of a definitive formulation of its nature and solutions.

3   *Various strategies are applied by different actors during environmental conflicts.* In the 10 cases of environmental conflicts examined in this book, citizens first tended to resort to formal channels, such as letters and visits, petitions, or appeals, to express their disagreement with government decisions. When formal channels were ineffective, they then chose informal strategies, such as group petitions, demonstrations, or even violent confrontations. Environmental NGOs sometimes provided suggestions and advice to local residents in order to show their disagreement with government decisions strategically, as happened in the Liulitun case. Experts applied various strategies in environmental conflicts: talking with government officials in a direct face-to-face way (the Xiamen case), providing professional suggestions (the Panyu case), or persuading residents using professional knowledge (the Wuxi case). Activists visited government agencies (the Liulitun case), organized collective activities to exert pressure on local governments (the Panyu case), facilitated the interactions of citizens and governments (the Songjiang case), or used social media to attract public attention (the Kunming case and the Dalian case). To sum up, diverse actors applied various strategies in order to influence the evolvement of environmental conflicts. This makes the evolution of environmental conflicts uncertain and unpredictable, ultimately posing a great challenge for local governments to govern them.

4   *Environmental conflicts evolve over time.* At the beginning of environmental conflicts, local governments are often busy expediting the approval of the proposed industrial projects, and the other actors do not know what is happening in the "black box." In the Wuxi case, local government secretly constructed a waste incineration power plant on the pretext of building a temple. Local residents did not know that a waste incineration power plant was being constructed until it started trial operation. During environmental conflicts, local governments mostly start to take the interests of other actors into account when protests occur or might occur. In the Dalian case and the Xiamen case, local municipalities stated that they would temporarily stop the PX plants when they realized that local residents planned to initiate large-scale protests. In the Panyu case, the protest initiated by local residents came as a surprise to local governments in Guangzhou, after which they claimed to have temporarily stopped the advancement of the waste incineration power plant. Sometimes, environmental conflicts end after protests, like the Ningbo case. In this case, the mobilized protesters attacked government officials; this greatly challenged

government authority, and the local government had to stop the debated project immediately. Sometimes, environmental conflicts evolve for a long time after protests, like in the Liulitun case. Local residents in Liulitun did not give up their opposition and continuously put pressure on Beijing local governments, which finally relocated the plant. In summation, it appears that no single actor can absolutely determine how environmental conflicts will evolve, as some uncertain conditions may arise and play a role in altering the evolutionary trajectory of environmental conflicts.

5   *Environmental conflicts are resolved in some respects, but not in a win–win way.* Win–win or zero–plus outcomes are established by governance scholars as the criteria by which to assess the substantive outcomes of environmental conflict resolution (Amy 1987; Bingham 1986; Fischer and Forester 1993; Glasbergen 1995; van Bueren, Klijn, and Koppenjan 2003). No new rounds of mobilized protests occurred at the end of the 10 cases; they thus were resolved from this perspective. In addition, five of the 10 debated industrial projects were relocated and two of them cancelled. One Beijing respondent (respondent 6) argued that relocating the Liulitun waste incineration power plant was a win–win outcome, as Beijing Municipality finally achieved its goal of constructing a waste incineration power plant and local residents obtained economic compensation. Meanwhile, the authoritative national mass media framed the Xiamen PX case and the Panyu case as the best practice for resolving environmental conflicts in China. The general comments were: local residents expressed their concerns in a rational way, and local governments actively interacted with citizens to respond to their concerns. Can these be viewed as win–win outcomes? The answer may be no. This depends on how benefits and costs are allocated among the actors involved and the extent to which the interaction processes are satisfactory for them.

In conclusion, the point of departure regarding the nature of Chinese environmental conflicts is that they are wicked problems occurring in a multi-actor context. Various actors, including local governments, central government, local residents, activists, experts, environmental NGOs, and private companies or SOEs with different perceptions about the nature and solution of environmental conflicts, apply diverse strategies that influence their evolvement.

What strategies are applied by local governments during environmental conflicts?

Regarding the strategies applied by Chinese local governments in addressing environmental conflicts, a typology consisting of six strategies was constructed in Chapter 3: go-alone, suppression, tension reduction, giving in, collaboration, and facilitation. This typology proved useful in identifying and categorizing the concrete government actions that emerged during environmental conflicts. Some key findings are presented as follows.

1   *Local governments ignore the interests of the other actors in the pre-phase of environmental conflicts.* Although the central government has put in place institutions, regulations, and policies to allow for public participation during the planning stage of industrial plants (such as EIAs), local governments often ignore or even manipulate them. At the beginning of the Tianjingwa case, Nanjing Municipality organized information disclosure during the period of the Chinese New Year. Few local residents knew this. One respondent (respondent 5) argued that local governments in Nanjing did not want to encourage public participation and it thus intentionally chose dates that were not convenient for public participation. When the other actors expressed their disagreement with local governments through formal channels, local governments often tended to have symbolical talks (the Liulitun case) or even used state force to repress them (the Tianjingwa case). It can be thus concluded that local governments tend to ignore the interests of the other actors in the project development stage.

2   *Local governments are reticent about using state force, but they often block information during environmental conflicts.* It might make sense for the state in authoritarian regimes to use state force to cope with protests initiated by citizens. To my surprise, it appears that Chinese local governments are comparatively reticent about using state force to cope with protests. Traditional state suppression occurred in two of the 10 cases: the Ningbo case and the Wuxi case. It seems that local governments tend to use state force to repress citizens when they think that their authority is significantly threatened and social order is in danger (the Ningbo case). They may also use it when it is too costly to make concessions (the Wuxi case). Furthermore, I established that information blockage is used to operationalize the suppression strategy. Compared to state repression, information blockage is more commonly used by local governments (such as in the Dalian, the Kunming, and the Pengzhou case).

3   *The giving in strategy is frequently used by local governments at the end of the environmental conflicts studied in this book.* Seven of the 10 cases of environmental conflicts ended in project relocation or project cancellation. This conclusion sounds counterintuitive, as it might generally be expected that the Chinese state would adhere to its own strategies. Unilateral project relocation implies a small concession on the part of local governments, but it does not imply that they always experience great losses, as their initial goals are accomplished. When local governments cancel debated projects, they generally experience significant losses, especially if they have made substantial financial investments in them, as in the Wuxi case. Apparently, local governments in China are not that effective in governing environmental conflicts.

4   *Collaboration and facilitation strategies do emerge, but they are relatively rare.* Collaboration and facilitation strategies emerged in two of the 10 cases: the Panyu case and the Xiamen case. In the Panyu case, local governments in Guangzhou organized symposiums to invite local residents, activists, experts,

and journalists to have a direct face-to-face discussion with the aim of building a consensus about how to resolve the waste problem in Guangzhou. In addition, Guangzhou Municipality established a commission that was specifically charged with providing consultancy to it regarding the resolution of the urban waste problem. This was an innovative institutional design to facilitate the governance of environmental conflicts. In the Xiamen case, Xiamen Municipality organized round-table meetings in which local governments, experts, and local residents had discussions with one another to jointly decide the future of the PX plant. These scenarios are not common in China though. In other cases, the collaboration and facilitation strategies were not adopted by local governments. In contrast to Western democracies, the Chinese government functions as the manager of the whole country. Civil society is very weak and has few opportunities to authentically shape government decisions. The Chinese state has got used to governing social affairs by itself. Negotiating with other actors for joint governance does not match its authoritarian nature. As for the facilitation strategy, it seems that the Chinese state uses it a lot in the economic field, but it rarely uses it in the governance of social conflicts. In this book, the application of the facilitation strategy by Guangzhou Municipality was indeed an innovative initiative.

5   *Most government strategies applied by local governments are still reactive rather than proactive.* Currently, Chinese local governments are still reactive in addressing environmental conflicts; this can be characterized as *regulation by escalations*, implying that they apply different strategies depending on the degree of escalation of opposition initiated by others (van Rooij 2012). In the short run, this is effective in maintaining social stability and state order; but it is not that helpful in fundamentally improving the effectiveness of governance in the long term. At the end of the 10 cases, no new rounds of protests occurred. However, citizens' trust in local governments was damaged in most cases and is unlikely to be remedied in an easy way. Environmental conflicts cannot be resolved solely through a top-down reactive approach. Rather, the Chinese state should design institutions to channel public participation, facilitate interaction, and empower citizens in order to reduce or prevent the occurrence of environmental conflicts. Promisingly, some proactive actions have occasionally been taken by Chinese local governments. At the end of the Tianjingwa case in Nanjing, for example, Nanjing Municipality established a consultation commission that was responsible for the risk assessment of mega industrial projects. In future, more studies about the performance of these proactive actions are needed.

6   *Local governments show variations regarding their application of strategies.* This book studied 10 cases of environmental conflicts that occurred in various cities of China. Some variations in terms of the application of government strategies can be identified. In general, local governments in Guangzhou did a relatively better job in governing environmental conflicts compared to the other cases. At the end of the Panyu case, they started to

create opportunities for mutual communication in order to build a consensus with the other actors with the aim of resolving the conflicts concerning the planning and construction of waste incineration power plants. This implies a widening of the scope regarding the resolution of environmental conflicts. Waste incineration was not the sole solution to resolve waste problem; waste reduction and waste re-usage were equally important. Local governments in other cases were struggling with the issue of where the facilities should be constructed with less anticipated resistance from local residents. To sum up, there are indeed variations across cases regarding government strategies in governing environmental conflicts: some local governments seem to act in a smarter way than others.

The above findings elaborate the application of government strategies during environmental conflicts; this is useful to enhance our understanding about how environmental conflicts are governed in China.

How can the application of government strategies during environmental conflicts be explained?

The third question relates to the explanation of why particular government strategies are applied during environmental conflicts. In Chapter 3, seven conditions were established upfront: the scale of protest, the form of protest, the position of higher-level governments, the position of the national mass media, the stage of projects, the involvement of activists, and the occurrence of events. Several key conclusions to answer the third question are drawn in the following.

1   *The position of higher-level governments is the most important condition in explaining the application of government strategies during environmental conflicts.* Proposition 5 in Chapter 3 suggested that, when higher-level governments criticize debated projects, local governments tend to apply a tension reduction, a giving in, a collaboration, or a facilitation strategy. However, when they support debated projects, local governments tend to adopt a go-alone or a suppression strategy. In Chapter 6, proposition 5 was both confirmed and disconfirmed: local governments may apply a go-alone strategy (confirmation) or a tension reduction strategy (disconfirmation) when higher-level governments show their support. In Chapter 7, proposition 5 was specified and reformulated: the support of higher-level governments for debated projects is important in explaining the second pattern of government strategies resulting in project continuation, and the absence of support from higher-level governments is important in explaining both the first pattern and third pattern of government strategies, respectively, resulting in project cancellation and project relocation. The csQCA analysis in Chapter 8 further specified proposition 5: when local governments refuse to make compromises, they always receive support from the central government. And when they make compromises, they never receive support from

the central government. Therefore, the position of higher-level governments is really a crucially important condition in explaining the application of government strategies.

2   *The form of protest is the second most important condition in explaining the application of government strategies during environmental conflicts.* In Chapter 3, I posited proposition 1: local governments tend to adopt a tension reduction or a giving in strategy when violent protests occur, whereas they tend to apply a go-alone or a suppression strategy when protests are peaceful. In Chapter 5 and Chapter 6, proposition 1 was disconfirmed and reformulated: local governments may apply a tension reduction or a giving in strategy to deal with the occurrence of peaceful protests. In Chapter 7, proposition 1 was specified and reformulated: the occurrence of violent protests is important in explaining project abandonment, and its absence is important in explaining project relocation and project continuation. This conclusion is not in line with what Cai (2010) concluded: that the occurrence of violent protests does not increase the odds of citizens being successful in achieving their goals. In this book, violent protests occurred in three cases, and local governments finally made compromises. For this reason, it seems that the occurrence of violent protests increases the odds of success for protesters. In Chapter 8, proposition 1 was further specified and reformulated: the occurrence and the absence of violent protests are two INUS conditions for governments to make compromises, and protests are always peaceful when local governments refuse to make compromises. In summary, the form of protest is one key condition in explaining the application of government strategies during environmental conflicts.

3   *The stage of projects is the third most important condition in explaining the application of government strategies during environmental conflicts.* Proposition 4 in Chapter 3 was: local governments tend to apply a tension reduction, a giving in, a collaboration, or a facilitation strategy when debated projects are in their planning stage, whereas they tend to adopt a go-alone or a suppression strategy when debated projects are in their late stage. In Chapter 5, this proposition was confirmed: local governments may adopt a giving in strategy when debated projects are in their planning stage. In Chapter 6, proposition 4 was disconfirmed and reformulated: local governments may apply a tension reduction strategy to cope with debated projects in their late stage. In Chapter 7, proposition 4 was specified and reformulated: the early stage of debated projects is important in explaining project relocation and the late stage of debated projects is important in explaining project continuation. However, it should be noted that this condition had a broad meaning: it refers to the (perceived) costs of project cancellation or project relocation. In Chapter 8, proposition 4 was further specified and reformulated: the stage of the project is an INUS condition in explaining the occurrence of government compromises. The early stage or the late stage of debated projects, combined with other conditions, can result in the occurrence of government compromises. When local governments

refuse to make compromises, debated plants are always in the late stage. To sum up, it can be concluded that the stage of the project is a crucial condition in explaining the application of government strategies in environmental conflicts.

4   *The position of the national mass media is important for the explanation of the application of government strategies during environmental conflicts.* Proposition 3 in Chapter 3 stated: when the national mass media criticize existing government strategies, local governments tend to apply a tension reduction, a giving in, a collaboration, or a facilitation strategy. However, when they support them, local governments tend to adopt a go-alone or a suppression strategy. In Chapter 7, proposition 3 was specified and reformulated: the absence of support from the national mass media is important in explaining project cancellation and project relocation during environmental conflicts. In Chapter 8, the position of the national government and the position of the national mass media were merged as one condition. Proposition 3 was further specified and reformulated: when local governments refuse to make compromises, they always receive support from the central government (including the national mass media). And when they make compromises, they never receive support from the central government (including the national mass media). In summary, the position of the national mass media is important in explaining the application of government strategies.

5   *The scale of protest matters to the application of (or shifts in) government strategies during environmental conflicts, but it is not a crucial condition.* Proposition 2 in Chapter 3 was that local governments tend to adopt a tension reduction or a giving in strategy when the scale of protest is large. However, local governments tend to apply a go-alone or a suppression strategy when the scale of protest is small. In Chapter 5, proposition 2 was disconfirmed and reformulated: local governments may adopt a tension reduction strategy to deal with small-scale protests. In Chapter 6, proposition 2 was confirmed: local governments apply a giving in strategy to deal with large-scale protests. In Chapter 7, proposition 2 was specified and reformulated: the scale of protest is not an important condition in explaining the similarities and differences in the patterns of government strategies in environmental conflicts. In Chapter 8, proposition 2 was further specified and reformulated: the scale of protest is an INUS condition for the occurrence of government compromises during environmental conflicts. Large-scale or small-scale protests, combined with the other conditions, can lead to government compromises. To sum up, the four empirical studies confirm that the scale of protest does influence the application of (or shifts in) government strategies, but it is not crucial. This conclusion is not in line with the conclusion drawn by Cai (2010), who argues the scale of protest plays a crucially important role in shaping government decisions in social conflicts.

6   *The involvement of activists does influence the application of (or shift in) government strategies, but its influence on the overall application of government strategies is limited.* Proposition 6 in Chapter 3 stated that, when

embedded activists are involved in environmental conflicts, local governments tend to apply a tension reduction, a giving in, a collaboration, or a facilitation strategy. If unembedded activists are involved, they tend to adopt a go-alone or a suppression strategy. Proposition 6 was both confirmed and disconfirmed in Chapter 5: local governments may apply various strategies to deal with the involvement of unembedded activists, such as a suppression (confirmation), or a collaboration and a facilitation strategy (disconfirmation). In Chapter 6, proposition 6 was disconfirmed and reformulated: local governments may apply a suppression strategy to deal with the involvement of embedded activists. The analysis in Chapter 7 specified proposition 6: the absence of activists is an important condition in explaining the pattern of government strategies that result in project cancellation. In short, all three empirical studies confirm that the involvement of activists influences shifts in government strategies. Nevertheless, it is not a crucial condition in explaining government strategies.

7   *The occurrence of events does influence shifts in government strategies, but its influence on the overall application of government strategies is limited.* Proposition 7 in Chapter 3 was: local governments tend to apply a tension reduction or a giving in strategy to cope with the occurrence of planned events. However, when unplanned events occur, the direction of the application of government strategies becomes uncertain. In Chapter 5, proposition 7 was confirmed: local governments may apply a tension reduction strategy to cope with the occurrence of a planned event. In Chapter 6, proposition 7 was specified and reformulated: local governments may apply a tension reduction strategy to cope with the occurrence of unplanned events. In Chapter 7, proposition 7 was further specified and reformulated: the absence of events is an important condition in explaining the pattern of government strategies that result in project cancellation. To sum up, the occurrence of events does influence shifts in government strategies. However, its influence on the explanation of the (overall) government strategies is limited.

8   *Some other conditions may also influence the application of government strategies in environmental conflicts.* In Chapter 5, it was found that the use of social media and the urgency of the waste problem influenced the application of government strategies during the Panyu waste incineration power plant case.

In sum, the above findings provide explanations for the application of government strategies during environmental conflicts. The seven propositions posited in Chapter 3 were confirmed, disconfirmed, or specified in the four empirical chapters. An overview of the propositions in this book is summarized in Table 9.1.

Table 9.1 Overview of the propositions in this book

Proposition in Chapter 3	Proposition in Chapter 5	Proposition in Chapter 6	Proposition in Chapter 7	Proposition in Chapter 8
1. Local governments tend to apply a go-alone or a suppression strategy to deal with peaceful protests, whereas they tend to apply a tension reduction or a giving in strategy to cope with violent protests.	**Disconfirmation:** Local governments may apply a tension reduction strategy to cope with peaceful protests.	**Disconfirmation:** Local governments may apply a giving in strategy to cope with peaceful protests.	**Specification:** The occurrence of violent protest is important in explaining project cancellation. Its absence is important in explaining project continuation and project relocation.	**Specification:** The occurrence or the absence of violent protests, combined with other conditions, can lead to the occurrence of government compromises. The absence of violent protests contributes to the absence of government compromises.
2. Local governments tend to adopt a tension reduction or a giving in strategy when the scale of protests is large. However, local governments tend to apply a go-alone or a suppression strategy when protests are small scale.	**Disconfirmation:** Local governments may adopt a tension reduction strategy to deal with small-scale protests.	**Confirmation:** Local governments may apply a giving in strategy to cope with large-scale protests.	**Specification:** The scale of protests is not crucial in explaining the patterns of government strategies.	**Specification:** Large-scale or small-scale protests, combined with other conditions, can result in government compromises.

3. When the national mass media criticize existing government strategies, local governments tend to apply a tension reduction, a giving in, a collaboration, or a facilitation strategy. However, when they support government strategies, local governments tend to adopt a go-alone or a suppression strategy.	—		*Specification:* The absence of support from national mass media is important in explaining project cancellation and project relocation.	*Specification:* The absence of support from central government (including the national mass media) contributes to government compromises. Its presence (including from the national mass media) contributes to the absence of government compromises.
4. Local governments tend to apply a tension reduction, a giving in, a collaboration, or a facilitation strategy when debated projects are in their early stage, whereas they tend to adopt a go-alone or a suppression strategy when debated projects are in their late stage	*Confirmation:* Local governments may apply a collaboration and a facilitation strategy when debated projects are in their planning stage.	*Disconfirmation:* Local governments may adopt a tension reduction strategy when debated projects are in their late stage.	*Specification:* The early stage of debated projects is important in explaining project relocation. The late stage of debated project is important in explaining project continuation.	*Specification:* The early stage or late stage of debated projects, together with other conditions, can result in government compromises. Their late stage contributes to the absence of government compromises.

continued

Table 9.1 Continued

Proposition in Chapter 3	Proposition in Chapter 5	Proposition in Chapter 6	Proposition in Chapter 7	Proposition in Chapter 8
5. Local governments tend to apply a tension reduction, a giving in, a collaboration, or a facilitation strategy when higher-level governments criticize debated projects, whereas they tend to apply a go-alone or a suppression strategy if higher-level governments support debated projects.	–	***Confirmation/ disconfirmation:*** Local governments may adopt a go-alone or a tension reduction strategy when higher-level governments support debated projects	***Specification:*** The support of higher-level governments is important in explaining project continuation, and the absence of their support is important in explaining project cancellation and project relocation.	***Specification:*** The absence of support from the central government contributes to government compromises. Its presence contributes to the absence of government compromises.
6. When embedded activists are involved in environmental conflicts, local governments tend to apply a tension reduction, a giving in, a collaboration, or a facilitation strategy. If unembedded activists are involved, they tend to adopt a go-alone or a suppression strategy.	***Confirmation/ disconfirmation:*** Local governments may apply a suppression, a collaboration, or a facilitation strategy to deal with the involvement of unembedded activists.	***Disconfirmation:*** Local governments may apply a suppression strategy to deal with the involvement of embedded activists.	***Specification:*** The absence of activists is an important condition in explaining the pattern of government strategies that result in project cancellation.	–

7. Local governments tend to apply a tension reduction strategy to cope with the occurrence of planned events. However, when unplanned events occur, the direction of the application of government strategies is uncertain.	**Confirmation:** Local governments apply a tension reduction strategy to cope with the occurrence of planned events.	**Specification:** Local governments may apply a tension reduction strategy to deal with the occurrence of unplanned events.
		Specification: The absence of events is an important condition in explaining the similarities in the government strategy pattern that results in project cancellation.

Contributions and reflections

In this section, the theoretical, practical, and methodological contributions made by this book, as well as reflections, are presented.

Contributions to, and reflections on, theories

This book makes two main theoretical contributions:

1   *Construction of a conceptual framework to describe and explain government strategies during Chinese environmental conflicts*: A conceptual framework was constructed in this book to inquire into the typologies of government strategies and their explanation. A conceptual framework is defined as an inquiry tool to determine key concepts and their relationships in order to explore the phenomena in which we are interested (Ostrom 2007). The first key concept is the *policy game*, which shows the game-like nature of environmental conflict resolution in the Chinese context. The second key concept is *government strategy*. A typology of government strategies was constructed to function as a heuristic tool to identify and categorize concrete government actions that emerge during environmental conflicts. One main contribution of this typology is that it demonstrates that conflict resolution can be a zero–plus game with the potential of achieving a win–win solution. This adds new building blocks – establishing the collaboration and the facilitation strategy as two optional choices for local governments – to Cai's (2010) typology of government strategies to resolve social conflicts. The third key concept is the conditions that explain the application of government strategies during environmental conflicts. Seven conditions were identified. Proposition and configurational thinking are used in this book to establish the relationships between these conditions and the application of government strategies during environmental conflicts. The former elucidates the causal relationships between the individual conditions and the application of government strategies. The latter inspired us to explore the causality between combinations of the conditions and the application of government strategies.

2   *Specification of the features of Chinese governance that contribute to theoretical debates that explain the China paradox*: In Chapter 1, I pointed out that some governance scholars are very interested in explaining the China paradox, referring to China's high economic development and improvement in social welfare but its relatively low score on government quality for all commonly used measures (Rothstein 2015). This book agreed with the arguments of some authors (like Tsang 2009) that the state's responsive nature plays an important role in explaining this. I first consolidated the responsive nature of the Chinese state: local governments smartly adjust their decisions in order to accommodate the demands of other actors. Therefore, it can be concluded that *responsive authoritarianism* exists in China (Cai 2004; van

Rooij, Stern, and Fürst 2016; Weller 2012). Additionally, this book further specifies the features of the responsiveness of Chinese governance. The responsiveness of the Chinese state is highly selective and conditional: it is not responsive to all issues. Local governments always prioritize responding to those issues in which they assume that the central government may intervene, or those that receive high attention from others (such as the mass media). In short, the Chinese state does not establish democracy, transparency, participation, or openness as its dominant values in social governance as governments in Western democracies do. The Chinese state is, however, very sensitive to various demands of citizens. It adjusts its strategies and policies to channel social grievances, although it never generously and easily meets its citizens' demands. This implies that citizens mostly have to make substantial efforts to anticipate an influence on government policies. This again illustrates the shortcoming of the responsiveness of Chinese governance: the responsiveness of Chinese governance is highly selective and mostly reaction-oriented. In short, the Chinese state's responsiveness may be an important cause of the China paradox. However, when environmental problems become more and more severe, this reactive response may increasingly become dysfunctional and threaten the basis of the China paradox: well-being for the people in return for party control.

Despite the theoretical contributions made by this book, one main theoretical issue remains: to what extent can the framework constructed in this book be applied in China and to what extent does the conceptual framework include Chinese particularities? Also, even after this study, the decision-making process in China is still largely an unopened black box. Some informal conditions might influence the decisions made by the Chinese state. One respondent interviewed in Beijing (respondent 11), for example, argued that local residents' personal networks (or *guanxi*) must play a role in relocating undesired projects. During my fieldwork in China, I found it very difficult to find respondents in public governance who would really speak freely and openly. More research is thus necessary to further open the black box regarding the classification of government strategies and the role of certain conditions that now remain a bit underexposed and hopefully may be more central in future research. However, it is noted that in all interviews (also in Western democracies) respondents' motives can only be constructed to a certain extent because we are not sure that respondents will tell us everything or whether they remember things correctly. By interviewing various respondents, and using different data sources (like news reports and documents), I have tried to compensate for this.

Contributions to, and reflections on, practices

This book offers practical insights for decision makers regarding the governance of environmental conflicts in China. Environmental conflicts pose a great challenge for the Chinese state given their presumed influence on social order.

Governing them in a more satisfactory way is a politically and practically relevant issue for governments. Some suggestions based on so-called best practices in Western democracies can be made. First, local governments should pay attention to the possibility of widening the scope for solving environmental conflicts. Take waste incineration power plants as an example. In contrast to establishing waste incineration as the only approach to resolve waste problems, more options could be adopted by local governments, such as institutionalizing waste classification, reducing waste production, and enhancing waste recycling. Second, local governments might want to resolve environmental conflicts in a more proactive way. In many instances, the occurrence of an environmental conflict comes as a surprise that shocks local governments. The reactive approach means that local governments have to work like firefighters to remedy earlier mismanagement. The reactive approach is ineffective in governing problems that are highly uncertain and unpredictable, such as the environmental conflicts studied in this book. A proactive approach is a better option. Local governments should involve stakeholders in various project phases from planning, through construction, to operation. Third, local governments could redesign the existing performance system to make government officials accountable for other social values, such as safety and social harmony. Fourth, local governments could learn to become qualified facilitators or mediators to know how to negotiate and collaborate with other actors. For example, local governments could learn to create platforms for public participation or design rules to enable mutual dialogues.

One reflection on practice that emanates from this book is that, despite China's authoritarian nature, in many cases the government has to make important concessions at a high cost, showing its ineffectiveness in dealing with environmental conflicts.

Contributions to, and reflections on, methods

Three case study strategies were applied in this book: a single case study, a comparative case study, and a QCA. Few researchers have used these three approaches in one study. Their combined use is one important methodological contribution from this book. The case study strategies all add new insights to the explanation of government strategies in environmental conflicts. In short, the single case study method revealed how the occurrence of, and shifts in, individual government strategies are explained during environmental conflicts. The comparative case study method allowed us to look at the 10 cases at a higher aggregation level and seek an explanation for the application of government strategies. QCA, like the comparative case study method, made it possible to study the 10 cases at a higher aggregation level. Also, it facilitated the study of how combinations of conditions shape the application of government strategies. In summary, all three case study strategies revealed different things and could add new insights to the explanation of government strategies in environmental conflicts. The use of the three methods allowed me to study government strategies during environmental conflicts using different approaches, based on which more valid and robust conclusions could be drawn. In the following, three limitations to the use of the three case study strategies are discussed.

1   *The dynamic dimensions of the cases are not well explored in the comparison of the cases.* Process tracing was applied in Chapters 5 and 6 to study why local governments adopted different strategies over time. The dynamic dimension of the cases was explored. However, the dynamic dimensions of the 10 cases did not get much attention in Chapters 7 and 8 in the comparative case study and the QCA. This is one limitation of this book. Some questions were not addressed: did the order of the conditions matter in anticipating different government strategies in the various cases? Does the time difference between the cases, such as the Xiamen case in 2007 and the Kunming case in 2013, matter in explaining the application of government strategies? Did earlier governance responses influence the choice of governance strategies later on? Did learning occur over time? Thus, more studies are needed in order to explore the dynamic dimensions of the cases.

2   *The generalizability of the three case study strategies could be further improved.* For all qualitative case studies, one unavoidable basic problem is *small-N, large number of variables.* Although this study combines three case study approaches, this does not solve the problem of the limited number of cases studied (see Lijphart 1971). This is the second limitation of this book. To what extent can the conclusions drawn in this book be applied to all the population (i.e. all types of environmental conflicts in China)? More studies are needed to answer this question.

3   *The empirical data and information could be enriched.* The third limitation relates to empirical data. Thirty-two interviews were conducted with the coordinators of environmental NGOs, government officials, experts, activists, and local residents. More empirical data are needed. In particular, more information collected from Chinese government officials through face-to-face interviews is necessary in order to further check the validity of the conclusions drawn in this book. However, this is challenging given the sensitivity of the issue – the governance of environmental conflicts.

A research agenda for studying the governance of environmental conflicts in China

At least three new research activities can be proposed as a follow-up to this study. They are presented below.

1   *It is important to conduct more case studies to analyze how (other types of) environmental conflicts are governed in China.* Ten cases of environmental conflicts were studied in this book. The number of cases studied is still limited, and more cases are needed to generalize the conclusions drawn in this book. Some scholars have already explored how other types of environmental conflicts are resolved in China. For example, Mertha (2009) has

studied how non-state actors succeed in changing government policies in relation to the fate of hydropower facilities in China. However, the study does not address the issue of government responses. It would be interesting for scholars to explore how other types of environmental conflicts in different sectors are governed in China, such as the debates concerning the construction of nuclear power plants, wind turbines, water management facilities, or high-speed railways.

2   *Quantitative methods could be used to test the findings of this book.* The combination of quantitative and qualitative methods is recommended in order to draw robust and valid conclusions (see Flyvbjerg 2006; Lieberman 2005). Quantitative methods (such as a survey) could be an option to test the propositions drafted in Chapter 3 in order to generalize the conclusions drawn in this book. Collecting data directly from government officials is difficult because local government officials mostly do not have any interest in becoming involved in social science research. More importantly, their involvement often requires permission from senior government officials, and this is not easy for researchers to access. One possibility might be to collect survey data from government officials who are trained at Party schools or universities. With empirical studies based on quantitative data, it would be possible to compare those analysis results with the propositions and conclusions drawn in this book to check their validity and generalizability.

3   *Other variants of QCA could be used to further study the explanation of government strategies in governing environmental conflicts.* At the end of Chapter 8, it was acknowledged that the use of csQCA requires researchers to dichotomize conditions and outcomes. However, social reality is not easily dichotomized. The use of csQCA thus posed challenges for me. One possible option to remedy this shortcoming is to use other variants of QCA, such as fuzzy-set QCA (fsQCA). Different from crisp-set, which shows mainly the qualitative dimension of the conditions or outcome, fuzzy set shows both the qualitative and the quantitative differences (Ragin 2008). It would be possible to calibrate the conditions and outcomes in membership scores from 0 to 1, allowing researchers to show the fine-grained variation across cases. In short, the conclusions drawn from fsQCA can be more precise because they tell us a lot about the relationships between the *degrees* of conditions and outcomes (Mahoney 2004). Also, it is possible for researchers to use Temporal QCA (TQCA) to explore how the time dimension plays a role in explaining the application of government strategies during environmental conflicts (Caren and Panofsky 2005).

Appendix 1

Description of the 10 cases

Case 1 The Xiamen PX case in 2007

In November 2006, Zhao Yufen, an academician working at Xiamen University noticed a news item in a local newspaper that announced that a PX project would be constructed soon in Xiamen. Together with other five experts, she afterwards sent a letter to Xiamen Municipality to express their worries about the negative influence of the proposed PX project on local residents' health. On December 6, they had a face-to-face conversation with the key leaders of Xiamen Municipality, but their opinions later failed to be treated seriously as the PX plant was further advanced. In March 2007, during the annual session of the National People's Political Consultative Conference (NPPCC) in Beijing, 105 representatives jointly submitted a proposal calling for the relocation of the Xiamen PX project. This proposal was regarded as the No. 1 proposal in the NPPCC. Afterwards, the National Development and Reform Commission dispatched a committee to conduct an on-site investigation into the PX plant, and the leader of the committee pointed out that the Xiamen PX project had met all the formal requirements and it was impossible to stop or relocate it. On May 27, 2007, construction of the PX project started. On the morning of May 30, the Executive Vice-Mayor of Xiamen Municipality stated that formal construction of the PX project had stopped. On June 1, 2007, between 8,000 and 10,000 citizens, wearing yellow armbands and holding banners, marched through the city to the Xiamen Municipality headquarters and demanded that Xiamen Municipality cancel the Xiamen PX project. After the protest, Xiamen Municipality determined to delay the construction of the PX plant. On June 7, Xiamen Municipality accepted the suggestion from the State Environmental Protection Administration to conduct an environmental impact assessment for the whole Haicang area, but Xiamen Municipality did not make any statement. On December 5, 2007, Xiamen Municipality and the Chinese Research Academy of Environmental Sciences jointly announced that the environmental impact assessment report for Haicang district was available; it concluded that Haicang district had limited physical space, and it must make a choice between a second city center and a petrochemical industry zone. On December 8, 2007, Xiamen Municipality opened an online voting system to sound out public opinion in order to decide the future of the PX project. It closed at 10:44 pm on December 9. The result indicated that 55,376 voters opposed the construction of the Xiamen PX project and that 3,078 voters supported it. Later, Xiamen Municipality claimed that there had been a technical flaw in the voting system and that the result was not accurate. On December 13 and 14, 2007, Xiamen Municipality organized two round-table discussions about the Xiamen PX project, after which the deputy mayor of Xiamen Municipality, who was in charge of the Xiamen PX project, went to the Fujian provincial government to report the actual situation about it. On December 19, an authoritative national newspaper, *Renmin Daily*, published an article that stated that the removal of the Xiamen PX project to another location was the best choice for Xiamen Municipality. At the end of December, the Fujian provincial government and Xiamen Municipality decided to relocate the PX project to the Gulei Peninsula.

Case 2 The Liulitun waste incineration power plant case in Beijing in 2007

On January 17, 2007, the Haidian Municipal Commission of City Administration organized a public forum, during which the invited experts in favor of waste incineration and local residents who opposed it had a hot debate on waste incineration. Subsequently, both the government bureau and the local residents felt disappointed. On January 30, 2007, the Beijing Environmental Protection Bureau approved the environmental impact assessment (EIA) report for the Liulitun waste incinerator. After that, local residents submitted two copies of the administrative review application separately to Beijing Municipality and the State Environmental Protection Administration (SEPA). On March 2, 2007, the Haidian District Municipal Commission of City Administration and Environment invited the local residents' representatives and the experts to organize a second public forum. However, it did not meet the locals' expectations in that it focused mainly on resolving the odor problem caused by the Liulitun waste landfill, making it look like a brainwashing meeting.

During the period of the National People's Congress and the National People's Political Consultative Conference (NPC and NPPCC, *Lianghui*) in March, 2007, Zhou Jinfeng, a representative of the NPPCC, submitted a proposal titled "On Stopping the Construction of Liulitun Waste Incineration Power Plant." On March 30, one national medium, *China Business Times* (*Zhonghua Gongshang Shibao*), reported the Liulitun case. In addition, several news programs, such as News 60 Minutes, produced by China Central Television (CCTV) broadcast the Liulitun case. On May 25, 2007, Beijing Municipality replied to the local residents' administrative review application that it agreed with the construction of the Liulitun waste incineration power plant. On the evening of June 4, some activists in Liulitun were informed that the key leaders of the Haidian district government would invite them to have a face-to-face conversation. On June 5, 2007, virtually 1,000 local residents wearing uniforms visited SEPA and stood outside of it, requiring to meet the SEPA Director. On that day, Wu

Yamei, the deputy mayor of the Haidian district government had a face-to-face conversation with the representatives of the participants involved in the protest. On June 7, Pan Yue, the Deputy Director of SEPA, publicly stated that construction of the Liulitun waste incineration power plant should be postponed.

On November 7, 2007, Beijing Municipality released a newly revised standard for air pollutants in Beijing titled "Emission Standard of Air Pollutants for Municipal Solid Waste." One revision was that the distance between waste incineration power plants and public facilities, such as residential communities, schools, and hospitals, must exceed 300 meters. This revision was regarded by some local residents as facilitating the construction of waste incineration power plants in Beijing. Beijing Municipality and the Haidian district government kept silent during the first half of 2008. From August 8 to 24, 2008, the Olympic Games were hosted in Beijing. On October 9, 2008, it was reported by the *Beijing Evening News* that the Liulitun waste incineration power plant project had finished its EIA. As local residents knew this, they drafted a petition letter with thousands of signatures and submitted it to the National People's Congress, the State Council, and other government agencies in Beijing. In March 2009, a spokesman on behalf of the Ministry of Environmental Protection restated that the Liulitun waste incineration power plant should be further discussed and not be constructed before it was approved by the Beijing Environmental Protection Bureau. On July 9, 2009, Zhao Lihua, the Deputy Director of Haidian District Municipal Commission of City Administration and Environment, had a face-to-face conversation with local residents and claimed that the Liulitun waste incineration power plant had been virtually halted.

On July 21, 2010, Beijing Municipality held a meeting to discuss the approval of a new waste incineration power plant project, the Lujiashan waste incineration power plant project. It completed all its approval procedures in three months. In addition, although it was a waste incineration power plant, it was renamed as Biomass Power Project of the Capital Steel Company in Beijing. On October 23, 2010, the Lujiashan waste incineration power plant project started formal construction. On November 16, 2010, the Haidian Environmental Protection Bureau organized a first EIA for another new waste incineration power plant project in Dagong village. On January 19, 2011, the Haidian district government determined the cancellation of the Liulitun waste incineration power plant. Meanwhile, Dagong village was established as the new alterative location.

Case 3 The Panyu waste incineration power plant case in Guangzhou in 2009 (see Chapter 5)

Case 4 The Tianjingwa waste incineration power plant case in Nanjing in 2009

In November 2008, the Nanjing City Landscape Bureau organized the second information disclosure for the Tianjingwa waste incineration power plant project.

From January 24 to February 11, 2009, during the period of the Chinese New Year, the third information disclosure was organized. Few people became aware of this, and the timing of the disclosure was regarded as a government tactic to impede public participation. On February 6, 2009, five days before the third information disclosure for the project, three representatives of local residents submitted their petition letter signed by 5,000 citizens to the Jiangsu Provincial Academy of Environmental Science, the environmental impact assessment (EIA) institute for the Tianjingwa waste incineration power plant. However, one staff member working there refused to receive their petition letter, stating that the EIA for the Tianjingwa waste incineration power plant project was completed and that the information disclosure was merely a formality.

On March 17, 2009, Nanjing Municipality organized a public hearing. Of the 20 representatives, 16 supported the construction of the plant. However, some local residents remarked that the public hearing was manipulated by Nanjing Municipality. After the public hearing, local mass media in Nanjing commented that most local residents were looking forward to the construction of the Tianjingwa waste incineration power plant as soon as possible. On March 27, 2009, over 20 local residents assembled outside the Jiangsu Provincial Environmental Protection Department in the hope of submitting their petition letter. One senior government official had a face-to-face conversation with five representatives of local residents. Some local residents questioned the legitimacy of the Tianjingwa waste incineration power plant, but the Nanjing Planning Bureau and Jiangsu Provincial Construction Department later respectively confirmed its legitimacy. On May 8, 2009, the Nanjing Urban Planning Bureau publicly stated that the project conformed to the national procedures. Subsequently, the Jiangsu Provincial Construction Department approved Tianjingwa as the new location for the construction of the plant on May 11, 2009.

Before the final approval of the Tianjingwa waste incineration power plant project, the Jiangsu Provincial Environmental Protection Department claimed that members of the public could voice their opinions from May 11 to 15, 2009, on whether the plant should be approved. On May 14, over 100 local residents assembled in front of the Jiangsu Provincial Environmental Protection Department, demanding a dialogue with its key leaders. At 3 pm, some persons came out of the government building, hustling and even hitting local residents. Some of the protesters had bleeding facial injuries, and some residents invited local mass media to report the action taken by the Jiangsu Provincial Environmental Protection Department; but no reporters came to the site. Then, local police went to the scene, and the violent confrontation was stopped. After this, some participants stood in the bus station near the Jiangsu Provincial Environmental Protection Department and sang the national anthem to attract public attention. Later, some government officials from Nanjing Municipality went there to persuade the local residents to return home. No public statements were released by the Jiangsu Provincial Environmental Protection Department, and the local residents returned home. On May 16, the Jiangsu Provincial Environmental Protection Department finally approved the construction of the Tianjingwa waste incineration power plant. Later, local residents applied to the Ministry for Environmental Protection for an administrative review; the ministry, however, supported the decision of the Jiangsu Provincial Environmental Protection Department.

On August 4, 2009, it was reported that the Tiangjingwa waste incineration power plant would soon be constructed. In September 2009, local residents in Tianjingwa applied to the Ministry of Housing and Urban-Rural Development (MHURD) for another administrative review. In November 2009, MHURD officials conducted an on-site investigation into the approval procedure for the Tianjingwa project. However, they did not reply to the administrative review requested by the local citizens.

In June, 2010, the deputy mayor of Nanjing Municipality publicly stressed that the Tianjingwa waste incineration power plant should be constructed as soon as possible. In 2014, the Youth Olympic Games would be hosted in Nanjing. Both Nanjing Municipality and the Jiangsu provincial government attempted to prioritize city landscape and environmental quality. In January 2011, it was reported that the Tianjingwa waste incineration power plant would be formally constructed and a waste classification policy would be implemented in Nanjing. In addition, it was proposed that 40 percent of urban waste would be classified before the Youth Olympic Games in 2014. At the beginning of 2012, Nanjing Municipality published a set of proposals during Twelfth Five-Year Plan Period. In it, two waste incineration power plants were planned to be constructed in Nanjing, one in Jiangning district, and the other in Xingdian town, Pukou district. Different from the earlier location of Tianjingwa, the new location for the construction of the waste incineration power plant was Xingdian, which had few nearby inhabitants.

Case 5 The Dalian PX case in 2011 (see Chapter 6)

Case 6 The Ningbo PX case in 2012

The chemical industries in the Ningbo Chemical Industry Zone in Zhenhai district caused serious environmental pollution in the local environment. Local residents had been expressing their complaints to the Zhenhai district government for several years. Their complaints, however, failed to be treated seriously. After announcing the construction of a paraxylene (PX) plant, the Zhenhai district government determined to move the chemical industries to the seaside and to move the nearby residential communities to inland of Ningbo. The resettlement of these residents required substantial amounts of money. Ningbo Municipality claimed that the Zhenhai refinery plant should contribute 10 billion Renminbi to resolve this. The Zhenhai refinery plant finally promised to contribute nine billion Renminbi. After obtaining financial support from the Zhenhai refinery plant, the Zhenhai district government started the resettlement project. It released a list of villages that would be removed. On October 22, 2012, over 200 residents went to the Zhenhai district government. Some of them were agitated and blocked a traffic junction. The Zhenhai district government organized a face-to-face conversation with the participants and promised that it would construct 20 residential communities to resettle them. On the morning of October 24, the Zhenhai district government released a statement in which it admitted the occurrence of a collective action initiated by local residents in Zhenhai district. Moreover, it reiterated that the construction of the PX plant met the national regulations.

Many local residents in Ningbo City center were shocked by the fact that a facility manufacturing PX would be constructed in Ningbo. They believed that the concentration of polluting factories in the Ningbo Chemical Industrial Zone had resulted in a surge of cancer. As a result, rumors were quickly disseminated through online forums in Ningbo. On October 25, some local residents in Zhenhai district blocked some main roads in order to show their opposition to the construction of the PX plant. The next day, on October 26, over 1,000 people assembled together in the center of Ningbo City. In the afternoon, the protesters assembled together, wearing masks and distributing pamphlets. Police officers blocked the roads to the city center in order to prevent more people from joining the rally. Moreover, local residents' microblog service did not work, making it impossible for them to send photos online. About 5 pm, over 100 local residents threw bricks and water bottles at the police officers. Between 7 pm and 8 pm, some protesters attacked a police car. At 11 pm, the police employed tear gas to disperse the local residents. During that day, 51 persons were detained. Among them, 13 people later received criminal convictions.

On the morning of October 27, over 1,000 local residents assembled in Ningbo City center. Some of them returned home after being persuaded by government officials. Some of them were arrested, but released several days later after lecturing and education by the police. The Zhenhai district government stated on that day that it would sound out public opinion and conduct an environmental impact assessment for the PX project. In the afternoon, Ningbo Municipality organized a meeting that the main leaders attended, and discussed the PX project. In the evening, Ningbo Municipality organized a public forum that the party secretary and the mayor of Ningbo Municipality attended. During this forum, local residents' representatives expressed their opinions about the PX project. Ningbo Municipality stated that it would further collect public opinions at the next stage to ensure that local residents' interests would be addressed. The same day, the Zhenhai Police Department released a statement in which it claimed that a small number of local residents attacked state agencies and warned that citizens should conform to the national laws to express their disagreement with government decisions. Furthermore, a local mass medium, *Ningbo Daily*, commented that the expression of grievances was allowed, but unlawful actions were not. On the evening of October 28, a news item was published by Zhenhai district government. It stated that the PX plant would not be constructed in Zhenhai.

Case 7 The Wuxi waste incineration power plant case in 2012

On March 18, 2011, Donggang town government organized a visit for local residents to the waste incineration power plants in Shanghai and Jiangyin. The Donggang town government originally aimed to convince local residents of the harmlessness of waste incineration. However, these residents actively conversed with the residents living near the two visited waste incineration power plants, and finally they found out that waste incineration was not as sophisticated as the town government had claimed. As a result, they became more worried about the negative influence of the waste incineration power plant on their health. In March 2011, some local residents built a wooden shed outside the gate of the Wuxi waste incineration power plant and some of them stayed there. On March 29 and 30, 2011, the Donggang town government organized a working group to persuade the local residents to return home. Subsequently, two residents were arrested, but released several days later. On April 2, 2011, Wuxi Municipality employed security guards to stay in the waste incineration power plant in order to ensure the facility's safety. On April 8, the Xishan district government organized a public forum in which two experts, Xiangxin Guo and Yuwen Ni, respectively from Beijing and Dalian, were invited to give a lecture on waste incineration. Thousands of local residents attended the public forum, but only about 200 of them were allowed to enter the public forum. Over 300 anti-riot police officers stood outside it. The invited experts left the public forum after their lecture. Local residents besieged the government officials. One senior government official from the Xishan district government promised that the Wuxi waste incineration power plant would not operate as long as local residents did not agree with it.

On April 12, 2011, the Xishan district government released a government instruction in which three issues were proposed: (1) the Wuxi waste incineration power plant project would be stopped until after the technical review organized by the Ministry of Environmental Protection (MEP); (2) the expert commission dispatched by the MEP would comprehensively review the Wuxi waste incineration power plant project; and (3) the Wuxi waste incineration power plant would not resume operations before the MEP made its final conclusion. On April 13, the Donggang town government stated that it would remove the shed constructed by the local residents. On April 15, the shed was removed. The local residents did not trust the experts invited by Wuxi Municipality and the Xishan district government, so they contacted some experts themselves. On May 5, 2011, the Xishan district government invited Zhangyuan Zhao from the Chinese Research Academy of Environmental Sciences to give a lecture titled "Waste Incineration and Public Health." Zhangyuan Zhao was a well-known expert who strongly opposed waste incineration. Over 400 residents attended this lecture. On May 8, the expert commission designated by the MEP started its on-site investigation and finally concluded that the Wuxi waste incineration power plant could start operation provided it further upgraded its facilities. The local residents were dissatisfied with this conclusion. Afterwards, they began to find ways to impede the operation of the Wuxi waste incineration power plant. With the help of the Center for Legal Assistance to Pollution Victims (CLAPV), local residents decided to litigate against the Jiangsu Environmental Protection Development. However, the court refused to handle this lawsuit because of time constraints.

On May 27, 2012, some old women went to the Donggang town government with the purpose of expressing their opposition to the operation of the Wuxi waste incineration power plant. On their way there, some of them were arrested by local police officers, and this caused great disappointment among local residents. The latter required the police officers to release the arrested persons, but the police officers refused to do so. As a result, a large-scale violent confrontation occurred. Over 100 residents got hurt, and tens of local residents were arrested. In addition, it was reported about 4,000 anti-riot police were dispatched to handle the protest. Afterwards, no media reported this case, and relevant information online was all deleted. Wuxi Municipality finally abandoned the waste incineration power plant.

Case 8 The Songjiang waste incineration power plant case in Shanghai in 2012

On May 17, 2012, the Songjiang Greening and Landscape Bureau announced that a new waste incinerator, with an investment of 0.25 billion Renminbi, would be constructed in Songjiang district. On May 18, 2012, the official microblog of the Songjiang district government announced this news. On May 22, 2012, the Songjiang Greening and City Landscape Management Bureau publicly announced this news. It determined that Songjiang district would construct a new waste incineration power plant, beside the existing waste landfill site. The Songjiang waste incineration power plant would dispose of 500 tons of waste per day. On the same day, the Shanghai Environmental Science Institute, the environmental impact assessment (EIA) institute, released the EIA report, and it also stated that public opinions would be sounded from May 22 to June 4.

On May 27, 2012, about 500 local residents in Songjiang took to the streets in order to attract the attention of the Songjiang district government and Shanghai Municipality. They assembled together and walked around the city center to express their opposition to the construction of the Songjiang waste incineration power plant. On May 31, the director of the Shanghai Greening and City Landscape Management Bureau commented in a local newspaper that waste incineration was an important way to resolve the odor problem in Songjiang. One week later, on June 2, 2012, one of the biggest commercial residential apartment projects held its opening ceremony in Songjiang. Some senior government officials of the Songjiang district government would attend this ceremony. When local residents learned this, they decided to go there. To avoid the occurrence of violent confrontations, three representatives of local residents met government officials, and they reached some agreements: (1) local residents would express their complaints in a rational and peaceful way without banners and slogans, and after the demonstration, they would leave the site as soon as possible; (2) local police would not harm and arrest local residents; and (3) senior government officials would have a direct face-to-face conversation with local residents. On June 2, about 600 local residents assembled together. However, this situation ultimately got out of control, resulting in a small-scale confrontation. On June 5, Tang Jiangfu, the general engineer of the Shanghai Greening and City Landscape Management Bureau, stated that waste incineration would be further advanced during the Twelfth Five-Year Plan Period (2010–2015). On June 8, 2012, the director of the Songjiang Greening and City Landscape Management Bureau apologized to local residents, and he defended the construction of the Songjiang waste incineration power plant, saying that it was an important strategy for local governments in Songjiang to address the odor problem caused by the Songjiang waste landfill. In addition, he suggested that collective "walking" initiated by local citizens was not the best way to express complaints.

On August 13, 2012, the Shanghai Environment Sciences Institute organized the second information disclosure for the Tianma waste incineration power plant. Its first EIA report was published on May 3, 2012. This implied that a new waste incineration power plant would be constructed in the conjunction area of Songjiang district and Qingpu district, Shanghai, and jointly constructed by the two districts. In future, it would be used to dispose of waste for the two districts. However, citizens from both districts were opposed to the plant, and they published online posts to oppose its construction. In December 2012, the EIA report for the Tianma waste incineration power plant was released by the Shanghai Environment Sciences Institute. Afterwards, Shanghai Municipality planned to organize a public hearing on the plant in January, 2013. However, this public hearing was criticized by local residents in Songjiang because they believed that the Songjiang district government manipulated the public hearing. The Tianma waste incineration power plant was to begin operation in 2015. On December 3, 2013, construction started.

Case 9 The Kunming refinery project in 2013

In January, 2013, the Kunming refinery project was formally approved by the National Development and Reform Commission. It would annually produce 10 million tons of terephthalic acid (TPA) and 6.5 million tons of paraxylene (PX). On March 25, 2013, a post in a microblog signed by Zheng Xiejian proposed that local residents in Kunming should oppose the construction of the PX project in Kunming. Later, Kunming Municipality set "Kunming PX Project" as sensitive words, and all related online posts were censored. On March 29, Kunming Municipality organized a press conference in which it announced that the industrial zone in Caopu, Anning district, Kunming City, was established as the location for the construction of the refinery project. On the same day, Green Kunming, a local voluntary organization, asked the Yunnan Provincial Environmental Protection Department for the environmental impact assessment (EIA) report for the Kunming PX project, but it did not receive any reply. Local residents in Kunming were worried about the potentially negative influence of the refinery project on their health because Kunming City center was located upwind of it. On May 4, 2013, about 2,000 local Kunming residents assembled together in a city-center square. Later, local police officers surrounded the square, and local residents were not allowed to enter it. During this protest, no violent action occurred. Some photos of the collective action were posted online, but were deleted soon. On the same day, the higher education institutes in Kunming received a notification from the Kunming Communist Party Commission, stating that no faculty or student was allowed to participate in the demonstration. On May 10, 2013, Kunming Municipality organized a press conference in which the mayor stated that, if most local residents disagreed with the construction of the PX project in Kunming, it would not be constructed. On May 16, 2013, more than 2,000 local residents gathered in front of the provincial government headquarters. This drew a large police presence and began with one arrest, but remained largely peaceful. At 4.30 pm, Li Wenrong, the mayor of Kunming Municipality, met the protesters. He claimed that he would personally supervise the EIA process for the refinery plant. Afterwards, the protesters gave up their protest.

On May 25, 2013, the Anning Industry and Commerce Bureau released a notification that local residents must register to buy masks. This notification was later widely criticized by many mass media in China. Finally, it was cancelled. On May 27, many regions in Kunming implemented the real-name registration regulation for printing. In addition, the white T-shirts that may be used as uniforms by local residents to organize collective actions were not allowed to be sold in Kunming. These regulations were criticized by some national mass media, such as *Guangming Daily* and *People's Daily* (or *Renmin Daily*). Five days later, the real-name registration regulation was called off. From June 6 to 10, 2013, the China–South Asia Expo, jointly hosted by the Yunnan provincial government and the Ministry of Commerce, was held in Kunming.

On June 24, a most important mouthpiece of the Chinese Communist Party, the *People's Daily*, commented that "PX is just a potential carcinogen like coffee that we drink every day." Moreover, the mass media in Kunming produced television programs showing the harmlessness of PX. On June 25, the EIA report for the Kunming refinery project was released and concluded that its environmental risks were acceptable. This report was attacked by local residents. The Kunming Environmental Protection Bureau promised to implement information disclosure. Afterwards, an environmental organization in Kunming, Public and Environmental Affairs, asked for the full EIA report for the refinery project and for the PX project. On the afternoon of June 25, the Kunming Environmental Protection Bureau provided the link to the EIA report of the refinery project and stated that there was no report for the PX project. In addition, Kunming Municipality established a one-month public participation in which local residents were allowed to review the EIA report for the Kunming refinery project. Experts were invited to consult with local residents. Meanwhile, formal construction of the Kunming refinery plant began. In September 2013, *Phoenix*, a Hong Kong media source, reported that the PX project in Kunming was to be constructed soon after the completion of the China–Myanmar oil pipeline. In addition, it was reported that outsiders were not allowed to conduct on-site investigations into the Kunming PX project.

Case 10 The Pengzhou PX plant in 2013

Pengzhou City is an administrative county-level (district-level) city in China. It is part of Chengdu City, Sichuan Province. The Pengzhou petrochemical project includes three mega projects, namely, a 10 million-ton oil refinery project, an 800,000-ton ethylene project, and a one million aromatics project. It was planned to be constructed by both the Chengdu Petrochemical Company and the China Petrochemical Company between 2005 and 2020. The Pengzhou PX plant is part of the Pengzhou petrochemical project. On April 20, 2013, an earthquake occurred in Sichuan Province. The Pengzhou petrochemical plant was constructed after the earthquake. Since the "420" earthquake, local residents in Pengzhou were worried about the potentially disastrous outcome for them if the plant was destroyed by an earthquake. To address these rumors, the Sichuan Petrochemical Corporation conducted a comprehensive review of the petrochemical plant and concluded that it had no safety problems. On April 29, 2013, Chengdu Municipality stated that it would conduct a project assessment for the Pengzhou petrochemical plant. To impede the formal operation of the Pengzhou petrochemical project, local residents planned to organize a demonstration in Chengdu City center on May 4. As the Fortune Global Forum was to be hosted in Chengdu from June 6 to 8, 2013, Chengdu Municipality realized that it should take action to prevent the demonstration. On May 3, the Chengdu Police Department stated that it would organize a combat exercise, which was regarded as an activity to prevent the demonstration on May 4 and 5, 2013.

Moreover, Chengdu Municipality rearranged the schedules of working days in Chengdu. It replaced Saturday and Sunday with Monday and Tuesday. This meant that citizens in Chengdu had to work on Saturday (May 4) and Sunday (May 5). On May 4 and 5, all students in Chengdu were required to stay at schools to prevent their involvement in the demonstration. On May 4, 2013, Tianfu Square in Chengdu City center was closed to visitors, and many police were on duty there. Moreover, the Chengdu Police Department demanded that, if any printing companies were asked to print anything related to the words, "health," "petrochemical project," "Pengzhou," or "PX," they must report it. Because of these measures, the local residents' planned large-scale demonstration did not occur. Afterwards, local residents expressed their disappointment in online forums. Many online posts criticizing the local governments in Pengzhou were widely forwarded.

On the evening of May 5, 2013, the official microblog of a national mass medium, *Renmin Daily*, released a post that claimed that Chengdu Municipality should resolve local residents' disagreements through deliberation and mutual dialogue. On May 8, *Renmin Daily* commented that large-scale collective action was not the best way for Chinese citizens to address their complaints. Afterwards, the Sichuan Petrochemical Corporation stated that it would establish a platform to facilitate communication among the public, entrepreneurs, and local governments. Moreover, it promised that it would organize on-site visits for local residents to the Pengzhou petrochemical plant. Thereafter, on-site visits were organized intensively. On May 10, over 40 representatives of citizens were invited to have an on-site visit to the Pengzhou petrochemical plant. On May 23, about 50 representatives visited it. From June 3 to 9, it was reported that about 700 representatives of Chengdu citizens visited the plant. On June 28, it was reported that over 50 local Chengdu residents had an on-site visit to it. In May 2013, China Central Television broadcast programs to inform the public of the harmless nature of PX. From June 6 to 8, the Fortune Global Forum was hosted in Chengdu. On June 24, *Renmin Daily* published a special column to describe the nature of PX, and many authoritative experts were invited to confirm its harmlessness. In August, 2013, *Chengdu Daily* published comments to explain the importance of the petrochemical industry to people's daily life, and it publicly responded to local residents' worries about the project's safety problem. Moreover, information about the Pengzhou petrochemical plant was provided, and some experts from various academic fields made positive comments about the Pengzhou petrochemical plant. In September, 2013, it was reported that the Pengzhou petrochemical plant would start trial operation in October 2013. Regarding the PX plant in Pengzhou, no information was provided.

Appendix 2

Interviewee list in China

No. of respondent	Location of interviewing	Personal information
1	Nanjing	Government official in the Pukou district government
2		Government official in the Environmental Protection Bureau in Pukou district
3		Manager of the Xingdian industrial park in Pukou district
4		Expert at Nanjing University
5		Government official in Jiangsu Provincial Environmental Protection Department
6	Beijing	Expert at Tsinghua University
7		Expert at National Central University in Taiwan
8		Coordinator in Nature University (NC), an environmental NGO
9		Coordinator in Friends of Nature (FON), an environmental NGO
10		Coordinator in Nature University (NU), an environmental NGO
11		Activist
12		Activist
13		Government official in the Haidian City Solid Waste Administration Bureau
14		Government official in the Haidian Planning Bureau in Beijing
15	Guangzhou	Expert at South China University of Technology
16		Expert at Sun-Yat Sen University
17		Activist
18		Expert at Sun-Yat Sen University
19		Activist
20		Activist
21	Dalian	Protester
22		Protester
23		Expert at Dalian University of Technology
24		Government official in Dalian Municipality
25		Policeman
26		Expert at Northeast University of Finance and Economics
27		Government official in Dalian Municipality
28		Government official in the office for petitions and appeals
29		Port expert in Dalian Port
30		Government official in Dalian Municipality
31		Government official in Dalian Municipality
32		Expert at Liaoning Normal University

References

Achen, Christopher H., and Duncan Snidal. 1989. "Rational Deterrence Theory and Comparative Case Studies." *World Politics* 41 (2): 143–69.

Ackerman, Bruce. 2004. *Deliberation Day*. New Haven, CT: Yale University Press.

Agranoff, Robert, and Michael McGuire. 2001. "Big Questions in Public Network Management Research." *Journal of Public Administration Research and Theory* 11 (3): 295–326.

Agranoff, Robert, and Michael McGuire. 2004. *Collaborative Public Management: New Strategies for Local Governments*. Washington, DC: Georgetown University Press.

Allison, Graham T. 1971. *Essence of Decision: Explaining the Cuban Missile Crisis*. Boston, MA: Little, Brown and Company.

Altshuler, Alan. 1965. "The Goals of Comprehensive Planning." *Journal of the American Institute of Planners* 31 (3): 186–95.

Amy, Douglas. 1983. "Environmental Mediation: An Alternative Approach to Policy Stalemates." *Policy Sciences* 15 (4): 345–65.

Amy, Douglas. 1987. *The Politics of Environmental Mediation*. New York: Columbia University Press.

Anderson, James E. 1984. *Public Policymaking*. New York: CBS College Publishing.

Ansell, Chris, and Alison Gash. 2008. "Collaborative Governance in Theory and Practice." *Journal of Public Administration Research and Theory* 18 (4): 543–71.

Ansfield, J. 2013. "Alchemy of a Protest: The Case of Xiamen PX." In *China and the Environment: The Green Revolution*, edited by S. Geall, 136–202. London: Zed Books.

Arnstein, Sherry R. 1969. "A Ladder of Citizen Participation." *Journal of the American Institute of Planners* 35 (4): 216–24.

Axelrod, Robert. 2006. *The Evolution of Cooperation*. New York: Basic Books.

Bacow, Lawrence, and Michael Wheeler. 1984. *Environmental Dispute Resolution*. New York: Springer.

Baumgartner, Frank, and Bryan D. Jones. 1993. *Agendas and Instability in American Politics*. 1st ed. Chicago: University of Chicago Press.

Baumgartner, Michael. 2013. "Detecting Causal Chains in Small-N Data." *Field Methods* 25 (1): 3–24.

Baumgartner, Michael. 2014. "Parsimony and Causality." *Quality & Quantity* 49 (2): 839–56.

Beach, Derek, and Rasmus Brun Pedersen. 2013. *Process-Tracing Methods: Foundations and Guidelines*. Ann Arbor, MI: University of Michigan Press.

Bennett, Andrew, and Colin Elman. 2006a. "Qualitative Research: Recent Developments in Case Study Methods." *Annual Review of Political Science* 9 (1): 455–76.

Bennett, Andrew, and Colin Elman. 2006b. "Complex Causal Relations and Case Study Methods: The Example of Path Dependence." *Political Analysis* 14 (3): 250–67.

Berry, Frances S., Ralph S. Brower, Sang Ok Choi, Wendy Xinfang Goa, HeeSoun Jang, Myungjung Kwon, and Jessica Word. 2004. "Three Traditions of Network Research: What the Public Management Research Agenda Can Learn from Other Research Communities." *Public Administration Review* 64 (5): 539–52.

Bie, Jiangbo, Martin de Jong, and Ben Derudder. 2015. "Greater Pearl River Delta: Historical Evolution towards a Global City-Region." *Journal of Urban Technology* 22 (2): 103–23.

Bingham, Gail. 1986. *Resolving Environmental Disputes: A Decade of Experience*. Washington, DC: Conservation Foundation.

Birkland, Thomas. 1997. *After Disaster: Agenda Setting, Public Policy, and Focusing Events*. Washington, DC: Georgetown University Press.

Birkland, Thomas. 1998. "Focusing Events, Mobilization, and Agenda Setting." *Journal of Public Policy* 18 (1): 53–74.

Bishop, Patrick, and Glyn Davis. 2002. "Mapping Public Participation in Policy Choices." *Australian Journal of Public Administration* 61 (1): 14–29.

Blatter, Joachim, and Markus Haverland. 2012. *Designing Case Studies: Explanatory Approaches in Small-N Research*. New York, NY: Palgrave Macmillan.

Blom-Hansen, Jens. 1997. "A 'New Institutional' Perspective on Policy Networks." *Public Administration* 75 (4): 669–93.

Bomberg, Elizabeth. 2007. "Policy Learning in an Enlarged European Union: Environmental NGOs and New Policy Instruments." *Journal of European Public Policy* 14 (2): 248–68.

Booher, David E., and Judith E. Innes. 2002. "Network Power in Collaborative Planning." *Journal of Planning Education and Research* 21 (3): 221–36.

Börzel, Tanja A. 1998. "Organizing Babylon – On the Different Conceptions of Policy Networks." *Public Administration* 76 (2): 253–73.

Bosley, P., and K. Bosley. 1988. "Public Acceptability of California's Wind Energy Developments: Three Studies." *Wind Engineering* 12 (5): 311–18.

Boswell, Terry, and Cliff Brown. 1999. "The Scope of General Theory Methods for Linking Deductive and Inductive Comparative History." *Sociological Methods & Research* 28 (2): 154–85.

Brady, Henry, and David Collier. 2010. *Rethinking Social Inquiry: Diverse Tools, Shared Standards*. 2nd ed. Lanham, MD: Rowman & Littlefield Publishers.

Braumoeller, Bear F., and Gary Goertz. 2000. "The Methodology of Necessary Conditions." *American Journal of Political Science* 44 (4): 844–58.

Brettell, Anna M. 2003. *The Politics of Public Participation and the Emergence of Environmental Proto-Movements in China*. Doctoral Dissertation. University of Maryland, College Park, MD.

Breunig, Christian. 2006. "The More Things Change, the More Things Stay the Same: A Comparative Analysis of Budget Punctuations." *Journal of European Public Policy* 13 (7): 1069–85.

Burke, Edmund M. 1968. "Citizen Participation Strategies." *Journal of the American Institute of Planners* 34 (5): 287–94.

Burningham, Kate. 2000. "Using the Language of NIMBY: A Topic for Research, Not an Activity for Researchers." *Local Environment* 5 (1): 55–67.

Byrne, David. 1998. *Complexity Theory and the Social Sciences: An Introduction*. London: Routledge.

Byrne, David. 2005. "Complexity, Configurations and Cases." *Theory, Culture & Society* 22 (5): 95–111.

Cai, Yongshun. 2002. "The Resistance of Chinese Laid-off Workers in the Reform Period." *The China Quarterly* 170: 327–44.

Cai, Yongshun. 2004. "Managed Participation in China." *Political Science Quarterly* 119 (3): 425–51.

Cai, Yongshun. 2008a. "Local Governments and the Suppression of Popular Resistance in China." *The China Quarterly* 193 (March): 24–42.

Cai, Yongshun. 2008b. "Power Structure and Regime Resilience: Contentious Politics in China." *British Journal of Political Science* 38 (3): 411–32.

Cai, Yongshun. 2008c. "Social Conflicts and Modes of Action in China." *The China Journal* 59 (January): 89–109.

Cai, Yongshun. 2010. *Collective Resistance in China: Why Popular Protests Succeed or Fail.* Stanford, CA: Stanford University Press.

Campbell, Donald. 1975. "III. 'Degrees of Freedom' and the Case Study." *Comparative Political Studies* 8 (2): 178–93.

Campbell, Marcia, and Donald Floyd. 1996. "Thinking Critically about Environmental Mediation." *Journal of Planning Literature* 10 (3): 235–47.

Caren, Neal, and Aaron Panofsky. 2005. "TQCA—A Technique for Adding Temporality to Qualitative Comparative Analysis." *Sociological Methods & Research* 34 (2): 147–72.

Chen, Xi. 2012. *Social Protest and Contentious Authoritarianism in China.* Cambridge: Cambridge University Press.

Cobb, Roger, and Charles Elder. 1983. *Participation in American Politics.* 2nd ed. Baltimore, MD: The Johns Hopkins University Press.

Cobb, Roger, Jennie-Keith Ross, and Marc Howard Ross. 1976. "Agenda Building as a Comparative Political Process." *The American Political Science Review* 70 (1): 126–38.

Collier, David. 2011. "Understanding Process Tracing." *PS: Political Science and Politics* 44 (4): 823–30.

Collier, Ruth Berins, and David Collier. 1991. *Shaping the Political Arena: Critical Junctures, the Labor Movement, and Regime Dynamics in Latin America.* Princeton, NJ: Princeton University Press.

Connor, Desmond M. 1988. "A New Ladder of Citizen Participation." *National Civic Review* 77 (3): 249–57.

Coppedge, Michael. 1999. "Thickening Thin Concepts and Theories: Combining Large N and Small in Comparative Politics." *Comparative Politics* 31 (4): 465–76.

Cronqvist, Lasse. 2017. *Tosmana—Tool for Small-N Analysis.* Version 1.54. Trier: University of Trier.

Cronqvist, Lasse, and Dirk Berg-Schlosser. 2009. "Multi-Value QCA (mvQCA)." In *Configurational Comparative Methods: Qualitative Comparative Analysis (QCA) and Related Techniques,* edited by Benoît Rihoux and Charles Ragin, 69–86. Thousand Oaks, CA: Sage.

Crozier, Michel, and Erhard Friedberg. 1980. *Actors and Systems: The Politics of Collective Action.* Chicago: University of Chicago Press.

Daniels, Steven, and Gregg Walker. 2001. *Working Through Environmental Conflict: The Collaborative Learning Approach.* Westport, CT: Praeger.

Day, J.C., and Thomas Gunton. 2003. "The Theory and Practice of Collaborative Planning in Resource and Environmental Management." *Environments* 31 (2): 5–19.

De Bruijn, Hans, and Ernst Ten Heuvelhof. 2008. *Management in Networks: On Multi-Actor Decision Making*. 1st ed. London: Routledge.

de Jong, Martin, Dong Wang, and Chang Yu. 2013. "Exploring the Relevance of the Eco-City Concept in China: The Case of Shenzhen Sino-Dutch Low Carbon City." *Journal of Urban Technology* 20 (1): 95–113.

Deng, Yanhua, and Guobin Yang. 2013. "Pollution and Protest in China: Environmental Mobilization in Context." *The China Quarterly* 214 (June): 321–36.

Dery, David. 1984. *Problem Definition in Policy Analysis*. Lawrence, KS: University Press of Kansas.

Dion, Douglas. 1998. "Evidence and Inference in the Comparative Case Study." *Comparative Politics* 30 (2): 127–45.

Dorius, Noah. 1993. "Land Use Negotiation Reducing Conflict and Creating Wanted Land Uses." *Journal of the American Planning Association* 59 (1): 101–6.

Dowding, Keith. 1995. "Model or Metaphor? A Critical Review of the Policy Network Approach." *Political Studies* 43 (1): 136–58.

Dryzek, John S. 1990. *Discursive Democracy: Politics, Policy, and Political Science*. Cambridge: Cambridge University Press.

Ducsik, Dennis W. 1981. "Citizen Participation in Power Plant Siting: Aladdin's Lamp or Pandora's Box?" *Journal of the American Planning Association* 47 (2): 154–66.

Dukes, E. Franklin. 2004. "What We Know about Environmental Conflict Resolution: An Analysis Based on Research." *Conflict Resolution Quarterly* 22 (1–2): 191–220.

Dul, Jan, Tony Hak, Gary Goertz, and Chris Voss. 2010. "Necessary Condition Hypotheses in Operations Management." *International Journal of Operations & Production Management* 30 (11): 1170–90.

Edelenbos, Jurian. 2005. "Institutional Implications of Interactive Governance: Insights from Dutch Practice." *Governance* 18 (1): 111–34.

Edelenbos, Jurian, and Erik-Hans Klijn. 2006. "Managing Stakeholder Involvement in Decision Making: A Comparative Analysis of Six Interactive Processes in the Netherlands." *Journal of Public Administration Research and Theory* 16 (3): 417–46.

Eisenhardt, Kathleen, and Melissa Graebner. 2007. "Theory Building from Cases: Opportunities and Challenges." *The Academy of Management Journal* 50 (1): 25–32.

Emerson, Kirk, Tina Nabatchi, and Stephen Balogh. 2012. "An Integrative Framework for Collaborative Governance." *Journal of Public Administration Research and Theory* 22 (1): 1–29.

Enserink, Bert, and René A.H. Monnikhof. 2003. "Information Management for Public Participation in Co-Design Processes: Evaluation of a Dutch Example." *Journal of Environmental Planning and Management* 46 (3): 315–44.

Felstiner, William L.F., Richard L. Abel, and Austin Sarat. 1980. "The Emergence and Transformation of Disputes: Naming, Blaming, Claiming..." *Law & Society Review* 15 (3/4): 631–54.

Fischer, Frank. 1990. *Technocracy and the Politics of Expertise*. Newbury Park, CA: Sage.

Fischer, Frank. 1993. "Citizen Participation and the Democratization of Policy Expertise: From Theoretical Inquiry to Practical Cases." *Policy Sciences* 26 (3): 165–87.

Fischer, Frank. 2000. *Citizens, Experts, and the Environment: The Politics of Local Knowledge*. Durham, NC: Duke University Press.

Fischer, Frank, and John Forester, eds. 1993. *The Argumentative Turn in Policy Analysis and Planning*. Durham, NC: Duke University Press.

Fischer, Frank, and Herbert Gottweis, eds. 2012. *The Argumentative Turn Revisited: Public Policy as Communicative Practice*. Durham, NC: Duke University Press.

Fisher, Roger, William L. Ury, and Bruce Patton. 2011. *Getting to Yes: Negotiating Agreement Without Giving In*. New York: Penguin.

Flyvbjerg, Bent. 2006. "Five Misunderstandings about Case-Study Research." *Qualitative Inquiry* 12 (2): 219–45.

Forester, John. 1987. "Planning in the Face of Conflict: Negotiation and Mediation Strategies in Local Land Use Regulation." *Journal of the American Planning Association* 53 (3): 303–14.

Forester, John. 1989. *Planning in the Face of Power*. Berkeley, CA: University of California Press.

Forester, John. 1999. *The Deliberative Practitioner: Encouraging Participatory Planning Processes*. Cambridge, MA: The MIT Press.

Forester, John. 2006. "Making Participation Work When Interests Conflict: Moving from Facilitating Dialogue and Moderating Debate to Mediating Negotiations." *Journal of the American Planning Association* 72 (4): 447–56.

Frame, Tanis M., Thomas Gunton, and J.C. Day. 2004. "The Role of Collaboration in Environmental Management: An Evaluation of Land and Resource Planning in British Columbia." *Journal of Environmental Planning and Management* 47 (1): 59–82.

Frendreis, John P. 1983. "Explanation of Variation and Detection of Covariation: The Purpose and Logic of Comparative Analysis." *Comparative Political Studies* 16 (2): 255–72.

Fukuyama, Francis. 2013. "What Is Governance?" *Governance* 26 (3): 347–68.

Fung, Archon. 2005. "Deliberation before the Revolution Toward an Ethics of Deliberative Democracy in an Unjust World." *Political Theory* 33 (3): 397–419.

Fung, Archon. 2006. "Varieties of Participation in Complex Governance." *Public Administration Review* 66 (s1): 66–75.

Fung, Archon, and Erik Wright, eds. 2003. *Deepening Democracy: Institutional Innovations in Empowered Participatory Governance*. London: Verso.

Gage, Robert, and Myrna Mandell, eds. 1990. *Strategies for Managing Intergovernmental Policies and Networks*. New York: Praeger.

Geddes, Barbara. 1990. "How the Cases You Choose Affect the Answers You Get: Selection Bias in Comparative Politics." *Political Analysis* 2 (1): 131–50.

George, Alexander L., and Andrew Bennett. 2005. *Case Studies and Theory Development in the Social Sciences*. 4th ed. Cambridge, MA: The MIT Press.

Gerring, John. 2007. *Case Study Research: Principles and Practices*. 1st ed. New York: Cambridge University Press.

Gerrits, Lasse, and Stefan Verweij. 2013. "Critical Realism as a Meta-Framework for Understanding the Relationships Between Complexity and Qualitative Comparative Analysis." *Journal of Critical Realism* 12 (2): 166–82.

Glasbergen, Pieter. 1995. *Managing Environmental Disputes: Network Management as an Alternative*. Dordrecht: Kluwer Academic Publishers.

Glavovic, Bruce C., E. Franklin Dukes, and Jana M. Lynott. 1997. "Training and Educating Environmental Mediators: Lessons from Experience in the United States." *Mediation Quarterly* 14 (4): 269–92.

Goelz, Darcey J. 2009. "China's Environmental Problems: Is a Specialized Court the Solution?" *Pacific Rim Law & Policy Journal* 18: 155–88.

Goertz, Gary, and Harvey Starr, eds. 2002. *Necessary Conditions: Theory, Methodology, and Applications*. Lanham, MD: Rowman & Littlefield Publishers.

Goertz, Gary, and James Mahoney. 2005. "Two-Level Theories and Fuzzy-Set Analysis." *Sociological Methods & Research* 33 (4): 497–538.

Goertz, Gary, and James Mahoney. 2012. *A Tale of Two Cultures: Qualitative and Quantitative Research in the Social Sciences*. Princeton, NJ: Princeton University Press.

Goertz, Gary, and James Mahoney. 2013. "Methodological Rorschach Tests Contrasting Interpretations in Qualitative and Quantitative Research." *Comparative Political Studies* 46 (2): 236–51.

Grabosky, Peter. 2013. "Beyond Responsive Regulation: The Expanding Role of Non-State Actors in the Regulatory Process." *Regulation & Governance* 7 (1): 114–23.

Gray, Barbara. 1985. "Conditions Facilitating Interorganizational Collaboration." *Human Relations* 38 (10): 911–36.

Gray, Barbara. 1989. *Collaborating: Finding Common Ground for Multiparty Problems*. 1st ed. San Francisco, CA: Jossey-Bass.

Gray, Barbara, and Donna J. Wood. 1991. "Collaborative Alliances: Moving from Practice to Theory." *The Journal of Applied Behavioral Science* 27 (1): 3–22.

Green-Pedersen, Christoffer, and John Wilkerson. 2006. "How Agenda-Setting Attributes Shape Politics: Basic Dilemmas, Problem Attention and Health Politics Developments in Denmark and the US." *Journal of European Public Policy* 13 (7): 1039–52.

Gregory, Robin. 2000. "Using Stakeholder Values to Make Smarter Environmental Decisions." *Environment: Science and Policy for Sustainable Development* 42 (5): 34–44.

Gregory, Robin, Tim McDaniels, and Daryl Fields. 2001. "Decision Aiding, Not Dispute Resolution: Creating Insights through Structured Environmental Decisions." *Journal of Policy Analysis and Management* 20 (3): 415–32.

Groeneveld, Sandra, Lars Tummers, Babette Bronkhorst, Tanachia Ashikali, and Sandra van Thiel. 2015. "Quantitative Methods in Public Administration: Their Use and Development Through Time." *International Public Management Journal* 18 (1): 61–86.

Hall, Peter. 1993. "Policy Paradigms, Social Learning, and the State: The Case of Economic Policymaking in Britain." *Comparative Politics* 25 (3): 275–96.

Hall, Peter. 2006. "Systematic Process Analysis: When and How to Use It." *European Management Review* 3 (1): 24–31.

Hall, Peter. 2013. "Tracing the Progress of Process Tracing." *European Political Science* 12 (1): 20–30.

Hammond, Daniel R. 2013. "Policy Entrepreneurship in China's Response to Urban Poverty." *Policy Studies Journal* 41 (1): 119–46.

Han, Heejin. 2013. "China's Policymaking in Transition: A Hydropower Development Case." *The Journal of Environment & Development* 22 (3): 313–36.

He, Guizhen, Arthur Mol, and Yonglong Lu. 2016. "Public protests against the Beijing-Shenyang high-speed railway in China." *Transportation Research Part D* 43: 1–16.

Heclo, Hugh. 1974. *Modern Social Politics in Britain and Sweden: From Relief to Income Maintenance*. 1st ed. New Haven, CT: Yale University Press.

Heiman, Michael. 1990. "From 'Not In My Backyard!' to 'Not In Anybody's Backyard!' Grassroots Challenges to Hazardous Waste Facility Siting." *Journal of the American Planning Association* 56 (3): 359–62.

Hicks, Alexander. 1994. "Qualitative Comparative Analysis and Analytical Induction: The Case of the Emergence of the Social Security State." *Sociological Methods & Research* 23 (1): 86–113.

Hisschemöller, Matthijs, and Rob Hoppe. 1995. "Coping with Intractable Controversies: The Case for Problem Structuring in Policy Design and Analysis." *Knowledge and Policy* 8 (4): 40–60.

Ho, Peter. 2001. "Greening without Conflict? Environmentalism, NGOs and Civil Society in China." *Development and Change* 32 (5): 893–921.

Ho, Peter, and Richard Edmonds, eds. 2007a. *China's Embedded Activism: Opportunities and Constraints of a Social Movement*. London: Routledge.

Ho, Peter, and Richard Edmonds, eds. 2007b. "Perspectives of Time and Change: Rethinking Embedded Environmental Activism in China." *China Information* 21 (2): 331–44.

Hogan, John, and David Doyle. 2007. "The Importance of Ideas: An A Priori Critical Juncture Framework." *Canadian Journal of Political Science* 40 (4): 883–910.

Hogwood, Brian W., and Lewis A. Gunn. 1984. *Policy Analysis for the Real World*. Oxford: Oxford University Press.

Homer-Dixon, Thomas. 1999. *Environment, Scarcity, and Violence*. Princeton, NJ: Princeton University Press.

Hoppe, Robert. 2002. "Cultures of Public Policy Problems." *Journal of Comparative Policy Analysis* 4 (3): 305–26.

Hoppe, Robert. 2011a. *The Governance of Problems: Puzzling, Powering and Participation*. Bristol: Policy Press.

Hoppe, Robert. 2011b. "Institutional Constraints and Practical Problems in Deliberative and Participatory Policy Making." *Policy & Politics* 39 (2): 163–86.

Hou, Yu, and Tianzhu Zhang. 2009. "Developing Fears: Environmental Conflicts and Pollution Accidents in China." *The Newsletter: Encouraging Knowledge and Enhancing the Study of Asia* 50: 12–13.

Hunter, Susan, and Kevin M. Leyden. 1995. "Beyond NIMBY: Explaining Opposition to Hazardous Waste Facilities." *Policy Studies Journal* 23 (4): 601–19.

Huxham, Chris. 2003. "Theorizing Collaboration Practice." *Public Management Review* 5 (3): 401–23.

Huxham, Chris, and Siv Vangen. 2000. "Ambiguity, Complexity and Dynamics in the Membership of Collaboration." *Human Relations* 53 (6): 771–806.

Huys, Menno, and Joop Koppenjan. 2009. "Policy Networks in Practice: The Debates on the Future of Amsterdam Airport Schiphol." In *The New Public Governance: Emerging Perspectives on the Theory and Practice of Public Governance*, edited by S. Osborne, 365–93. New York: Routledge.

Hysing, Erik. 2009a. "From Government to Governance? A Comparison of Environmental Governing in Swedish Forestry and Transport." *Governance* 22 (4): 647–72.

Hysing, Erik. 2009b. "Governing Without Government? The Private Governance of Forest Certification in Sweden." *Public Administration* 87 (2): 312–26.

Imperial, Mark T. 2005. "Using Collaboration as a Governance Strategy: Lessons from Six Watershed Management Programs." *Administration & Society* 37 (3): 281–320.

Innes, Judith E. 1996. "Planning through Consensus Building: A New View of the Comprehensive Planning Ideal." *Journal of the American Planning Association* 62 (4): 460–72.

Innes, Judith E. 2004. "Consensus Building: Clarifications for the Critics." *Planning Theory* 3 (1): 5–20.

Innes, Judith E., and David E. Booher. 1999a. "Consensus Building and Complex Adaptive Systems: A Framework for Evaluating Collaborative Planning." *Journal of the American Planning Association* 65 (4): 412–23.

Innes, Judith E., and David E. Booher. 1999b. "Consensus Building as Role Playing and Bricolage: Toward a Theory of Collaborative Planning." *Journal of the American Planning Association* 65 (1): 9–26.

Innes, Judith E., and David E. Booher. 2004. "Reframing Public Participation: Strategies for the 21st Century." *Planning Theory & Practice* 5 (4): 419–36.

Isett, Kimberley R., Ines A. Mergel, Kelly LeRoux, Pamela A. Mischen, and R. Karl Rethemeyer. 2011. "Networks in Public Administration Scholarship: Understanding Where We Are and Where We Need to Go." *Journal of Public Administration Research and Theory* 21 (s1): i157–73.

Jervis, Robert. 1989. "Rational Deterrence: Theory and Evidence." *World Politics* 41 (2): 183–207.

Jessop, Bob. 1998. "The Rise of Governance and the Risks of Failure: The Case of Economic Development." *International Social Science Journal* 50 (155): 29–45.

John, Peter. 1999. "Ideas and Interests: Agendas and Implementation: An Evolutionary Explanation of Policy Change in British Local Government Finance." *The British Journal of Politics & International Relations* 1 (1): 39–62.

John, Peter. 2006. "Explaining Policy Change: The Impact of the Media, Public Opinion and Political Violence on Urban Budgets in England." *Journal of European Public Policy* 13 (7): 1053–68.

Johnson, Thomas. 2008. "New Opportunities, Same Constraints: Environmental Protection and China's New Development Path." *Politics* 28 (2): 93–102.

Johnson, Thomas. 2010. "Environmentalism and NIMBYism in China: Promoting a Rules-Based Approach to Public Participation." *Environmental Politics* 19 (3): 430–48.

Johnson, Thomas. 2013a. "The Health Factor in Anti-Waste Incinerator Campaigns in Beijing and Guangzhou." *The China Quarterly* 214 (June): 356–75.

Johnson, Thomas. 2013b. "The Politics of Waste Incineration in Beijing: The Limits of a Top-Down Approach?" *Journal of Environmental Policy & Planning* 15 (1): 109–28.

Johnson, Thomas. 2014. "Good Governance for Environmental Protection in China: Instrumentation, Strategic Interactions and Unintended Consequences." *Journal of Contemporary Asia* 44 (2): 241–58.

Johnson, Thomas. 2016. "Regulatory Dynamism of Environmental Mobilization in Urban China." *Regulation & Governance* 10 (1): 14–28.

Jones, Bryan D., and Frank Baumgartner. 2004. "Representation and Agenda Setting." *Policy Studies Journal* 32 (1): 1–24.

Jones, Bryan D., and Frank Baumgartner. 2005. *The Politics of Attention: How Government Prioritizes Problems*. Chicago: University of Chicago Press.

Jones, Bryan D., Frank Baumgartner, and James L. True. 1998. "Policy Punctuations: U.S. Budget Authority, 1947–1995." *The Journal of Politics* 60 (1): 1–33.

Jordan, Grant. 1990. "Sub-Governments, Policy Communities and Networks Refilling the Old Bottles?" *Journal of Theoretical Politics* 2 (3): 319–38.

Kay, Adrian, and Phillip Baker. 2015. "What Can Causal Process Tracing Offer to Policy Studies? A Review of the Literature." *Policy Studies Journal* 43 (1): 1–21.

Kickert, Walter J.M., Erik-Hans Klijn, and Joop Koppenjan. 1997. *Managing Complex Networks: Strategies for the Public Sector*. London: Sage.

King, Cheryl, Kathryn Feltey, and Bridget Susel. 1998. "The Question of Participation: Toward Authentic Public Participation in Public Administration." *Public Administration Review* 58 (4): 317–26.

King, Gary, Robert O. Keohane, and Sidney Verba. 1994. *Designing Social Inquiry: Scientific Inference in Qualitative Research*. Princeton, NJ: Princeton University Press.

Kingdon, John W. 2010. *Agendas, Alternatives, and Public Policies*. 2nd ed. Boston, MA: Pearson.

Klijn, Erik-Hans. 2001. "Rules as Institutional Context for Decision Making in Networks: The Approach to Postwar Housing Districts in Two Cities." *Administration & Society* 33 (2): 133–64.

Klijn, Erik-Hans, and Joop Koppenjan. 2000a. "Politicians and Interactive Decision Making: Institutional Spoilsports or Playmakers." *Public Administration* 78 (2): 365–87.

Klijn, Erik-Hans, and Joop Koppenjan. 2000b. "Public Management and Policy Networks: Foundations of a Network Approach to Governance." *Public Management an International Journal of Research and Theory* 2 (2): 135–58.

Klijn, Erik-Hans, and Joop Koppenjan. 2006. "Institutional Design." *Public Management Review* 8 (1): 141–60.

Klijn, Erik-Hans, and Joop Koppenjan. 2012. "Governance Network Theory: Past, Present and Future." *Policy & Politics* 40 (4): 587–606.

Klijn, Erik-Hans, and Joop Koppenjan. 2016. *Governance Networks in the Public Sector*. Oxon: Routledge.

Klijn, Erik-Hans, Joop Koppenjan, and Katrien Termeer. 1995. "Managing Networks in the Public Sector: A Theoretical Study of Management Strategies in Policy Networks." *Public Administration* 73 (3): 437–54.

Klijn, Erik-Hans, and Chris Skelcher. 2007. "Democracy and Governance Networks: Compatible or Not?" *Public Administration* 85 (3): 587–608.

Klijn, Erik-Hans, and Geert R. Teisman. 1997. "Strategies and Games in Networks." In *Managing Complex Networks: Strategies for the Public Sector*, edited by Walter J.M. Kickert, Erik-Hans Klijn, and Joop Koppenjan, 98–118. London: Sage.

Klir, George J., Ute St Clair, and Bo Yuan. 1997. *Fuzzy Set Theory: Foundations and Applications*. 1st ed. Upper Saddle River, NJ: Prentice Hall.

Kooiman, Jan. 1993. *Modern Governance: New Government-Society Interactions*. London: Sage.

Kooiman, Jan. 2003. *Governing as Governance*. 1st ed. London: Sage.

Kooiman, Jan, and Svein Jentoft. 2009. "Meta-Governance: Values, Norms and Principles, and the Making of Hard Choices." *Public Administration* 87 (4): 818–36.

Koppenjan, Joop, Mirjam Kars, and Haiko van der Voort. 2009. "Vertical Politics in Horizontal Policy Networks: Framework Setting as Coupling Arrangement." *Policy Studies Journal* 37 (4): 769–92.

Koppenjan, Joop, and Erik-Hans Klijn. 2004. *Managing Uncertainties in Networks: Public Private Controversies*. 1st ed. New York: Routledge.

Kraft, Michael E., and Bruce B. Clary. 1991. "Citizen Participation and the Nimby Syndrome: Public Response to Radioactive Waste Disposal." *The Western Political Quarterly* 44 (2): 299–328.

Kübler, Daniel. 2001. "Understanding Policy Change with the Advocacy Coalition Framework: An Application to Swiss Drug Policy." *Journal of European Public Policy* 8 (4): 623–41.

Lang, Graeme, and Ying Xu. 2013. "Anti-Incinerator Campaigns and the Evolution of Protest Politics in China." *Environmental Politics* 22 (5): 832–48.

Leach, William D. 2006. "Collaborative Public Management and Democracy: Evidence from Western Watershed Partnerships." *Public Administration Review* 66 (s1): 100–10.

Leach, William D., Neil W. Pelkey, and Paul A. Sabatier. 2002. "Stakeholder Partnerships as Collaborative Policymaking: Evaluation Criteria Applied to Watershed Management in California and Washington." *Journal of Policy Analysis and Management* 21 (4): 645–70.

Lecy, Jesse D., Ines A. Mergel, and Hans Peter Schmitz. 2014. "Networks in Public Administration: Current Scholarship in Review." *Public Management Review* 16 (5): 643–65.

Ledermann, Simone. 2012. "Exploring the Necessary Conditions for Evaluation Use in Program Change." *American Journal of Evaluation* 33 (2): 159–78.

Lee, Yuen-Ching Bellette. 2013. "Global Capital, National Development and Transnational Environmental Activism: Conflict and the Three Gorges Dam." *Journal of Contemporary Asia* 43 (1): 102–26.

Lewis, Jenny M. 2011. "The Future of Network Governance Research: Strength in Diversity and Synthesis." *Public Administration* 89 (4): 1221–34.

Li, Wanxin, Jieyan Liu, and Duoduo Li. 2012. "Getting Their Voices Heard: Three Cases of Public Participation in Environmental Protection in China." *Journal of Environmental Management* 98 (1): 65–72.

Li, Yanwei. 2017. "Governing Environmental Conflicts in China: Lessons Learned from the Case of the Liulitun Waste Incineration Power Plant in Beijing." *Public Policy and Administration*, Doi: 10.1177/0952076717709521.

Li, Yanwei, Vincent Homburg, Martin De Jong, and Joop Koppenjan. 2016a. "Government Responses to Environmental Conflicts in Urban China: The Case of the Panyu Waste Incineration Power Plant in Guangzhou." *Journal of Cleaner Production* 134: 354–61.

Li, Yanwei, Joop Koppenjan, and Vincent Homburg. 2017. "Governing Environmental Conflicts: A Comparative Analysis of Ten Protests against Industrial Facilities in Urban China." *Local Government Studies* 43 (6): 992–1013.

Li, Yanwei, Stefan Verweij, and Joop Koppenjan. 2016b. "Governing Environmental Conflicts in China: Under What Conditions Do Local Governments Compromise?" *Public Administration* 94 (3): 806–22.

Lieberman, Evan S. 2005. "Nested Analysis as a Mixed-Method Strategy for Comparative Research." *The American Political Science Review* 99 (3): 435–52.

Lieberson, Stanley. 1987. *Making It Count: The Improvement of Social Research and Theory.* Berkeley, CA: University of California Press.

Lieberson, Stanley. 1991. "Small N's and Big Conclusions: An Examination of the Reasoning in Comparative Studies Based on a Small Number of Cases." *Social Forces* 70 (2): 307–20.

Lieberson, Stanley. 1994. "More on the Uneasy Case for Using Mill-Type Methods in Small-N Comparative Studies." *Social Forces* 72 (4): 1225–37.

Lijphart, Arend. 1971. "Comparative Politics and the Comparative Method." *The American Political Science Review* 65 (3): 682–93.

Lijphart, Arend. 1975. "II. The Comparable-Cases Strategy in Comparative Research." *Comparative Political Studies* 8 (2): 158–77.

Liu, Yi, Yanwei Li, Bao Xi, and Joop Koppenjan. 2016. "A Governance Network Perspective on Environmental Conflicts in China: Findings from the Dalian PX Conflict." *Policy Studies* 37 (4): 314–31.

Logsdon, Jeanne M. 1991a. "Interests and Interdependence in the Formation of Social Problem-Solving Collaborations." *The Journal of Applied Behavioral Science* 27 (1): 23–37.

Logsdon, Jeanne M. 1991b. "Collaboration to Regulate L.U.S.T.: Leaking Underground Storage Tanks in Silicon Valley." *Journal of Business Research* 23 (1): 99–111.

Lubell, Mark. 2004a. "Collaborative Environmental Institutions: All Talk and No Action?" *Journal of Policy Analysis and Management* 23 (3): 549–73.

Lubell, Mark. 2004b. "Collaborative Watershed Management: A View from the Grassroots." *Policy Studies Journal* 32 (3): 341–61.

Lubell, Mark. 2004c. 'Resolving Conflict and Building Cooperation in the National Estuary Program'. *Environmental Management* 33 (5): 677–91.

Lubell, Mark, Mark Schneider, John Scholz, and Mihriye Mete. 2002. "Watershed Part-nerships and the Emergence of Collective Action Institutions." *American Journal of Political Science* 46 (1): 148–63.

Lucas, Samuel R., and Alisa Szatrowski. 2014. "Qualitative Comparative Analysis in Critical Perspective." *Sociological Methodology* 44 (1): 1–79.

Mackie, John. 1980. *The Cement of the Universe: A Study of Causation*. Oxford: Claren-don Press.

Mahoney, James. 1999. "Nominal, Ordinal, and Narrative Appraisal in Macrocausal Ana-lysis." *American Journal of Sociology* 104 (4): 1154–96.

Mahoney, James. 2004. "Comparative-Historical Methodology." *Annual Review of Soci-ology* 30: 81–101.

Mahoney, James. 2007. "Qualitative Methodology and Comparative Politics." *Com-parative Political Studies* 40 (2): 122–44.

Majone, Giandomenico. 1992. *Evidence, Argument, and Persuasion in the Policy Process*. New Haven, CT: Yale University Press.

Mandell, Myrna. 2001. *Getting Results Through Collaboration: Networks and Network Structures for Public Policy and Management*. Westport, CT: Praeger.

Marin, Bernd, and Renate Mayntz, eds. 1991. *Policy Networks: Empirical Evidence and Theoretical Considerations*. Boulder, CO: Westview Press.

Marsh, David, and Rod Rhodes, eds. 1992. *Policy Networks in British Government*. 1st ed. Oxford: Clarendon Press.

Marx, Axel, Benoît Rihoux, and Charles Ragin. 2014. "The Origins, Development, and Application of Qualitative Comparative Analysis: The First 25 Years." *European Polit-ical Science Review* 6 (1): 115–42.

McAdam, Doug, and Hilary Boudet. 2012. *Putting Social Movements in Their Place: Explaining Opposition to Energy Projects in the United States, 2000–2005*. 1st ed. Cambridge: Cambridge University Press.

McAvoy, Gregory E. 1998. "Partisan Probing and Democratic Decision Making: Rethink-ing the Nimby Syndrome." *Policy Studies Journal* 26 (2): 274–92.

McAvoy, Gregory E. 1999. *Controlling Technocracy: Citizen Rationality and the Nimby Syndrome*. Washington, DC: Georgetown University Press.

McClymont, Katie, and Paul O'Hare. 2008. "'We're Not NIMBYs!' Contrasting Local Protest Groups with Idealised Conceptions of Sustainable Communities." *Local Environment* 13 (4): 321–35.

Mertha, Andrew. 2008. *China's Water Warriors: Citizen Action and Policy Change*. Ithaca, NY: Cornell University Press.

Mertha, Andrew. 2009. "'Fragmented Authoritarianism 2.0': Political Pluralization in the Chinese Policy Process." *The China Quarterly* 200: 995–1012.

Michelson, Ethan. 2007. "Climbing the Dispute Pagoda: Grievances and Appeals to the Official Justice System in Rural China." *American Sociological Review* 72 (3): 459–85.

Michelson, Ethan. 2008. "Justice from Above or Below? Popular Strategies for Resolving Grievances in Rural China." *The China Quarterly* 193 (March): 43–64.

Mill, John Stuart. 1843. *A System of Logic, Ratiocinative and Inductive: Being a Con-nected View of the Principles of Evidence and the Methods of Scientific Investigation*. New York: Harper & Brothers.

Mintrom, Michael, and Phillipa Norman. 2009. "Policy Entrepreneurship and Policy Change." *Policy Studies Journal* 37 (4): 649–67.

Mintrom, Michael, and Sandra Vergari. 1996. "Advocacy Coalitions, Policy Entrepren-eurs, and Policy Change." *Policy Studies Journal* 24 (3): 420–34.

Mintzberg, Henry. 1978. "Patterns in Strategy Formation." *Management Science* 24 (9): 934–48.

Mol, Arthur P. J. 2006. "Environment and Modernity in Transitional China: Frontiers of Ecological Modernization." *Development and Change* 37 (1): 29–56.

Moore, Barrington. 1993. *Social Origins of Dictatorship and Democracy: Lord and Peasant in the Making of the Modern World.* Boston, MA: Beacon Press.

Most, Benjamin A., and Harvey Starr. 1982. "Case Selection, Conceptualizations and Basic Logic in the Study of War." *American Journal of Political Science* 26 (4): 834–56.

Nabatchi, Tina. 2007. "The Institutionalization of Alternative Dispute Resolution in the Federal Government." *Public Administration Review* 67 (4): 646–61.

Nelson, David, and Susan Webb Yackee. 2012. "Lobbying Coalitions and Government Policy Change: An Analysis of Federal Agency Rulemaking." *The Journal of Politics* 74 (2): 339–53.

Newman, Peter, and Isabella Jennings. 2008. *Cities as Sustainable Ecosystems: Principles and Practices.* Washington, DC: Island Press.

Nohrstedt, Daniel. 2005. "External Shocks and Policy Change: Three Mile Island and Swedish Nuclear Energy Policy." *Journal of European Public Policy* 12 (6): 1041–59.

Nohrstedt, Daniel. 2008. "The Politics of Crisis Policymaking: Chernobyl and Swedish Nuclear Energy Policy." *Policy Studies Journal* 36 (2): 257–78.

Nohrstedt, Daniel. 2010. "Do Advocacy Coalitions Matter? Crisis and Change in Swedish Nuclear Energy Policy." *Journal of Public Administration Research and Theory* 20 (2): 309–33.

Nohrstedt, Daniel. 2011. "Shifting Resources and Venues Producing Policy Change in Contested Subsystems: A Case Study of Swedish Signals Intelligence Policy." *Policy Studies Journal* 39 (3): 461–84.

Nohrstedt, Daniel, and Christopher Weible. 2010. "The Logic of Policy Change after Crisis: Proximity and Subsystem Interaction." *Risk, Hazards & Crisis in Public Policy* 1 (2): 1–32.

O'Brien, Kevin J. 1996. "Rightful Resistance." *World Politics* 49 (1): 31–55.

O'Brien, Kevin J., and Yanhua Deng. 2015. "Repression Backfires: Tactical Radicalization and Protest Spectacle in Rural China." *Journal of Contemporary China* 24 (93): 457–70.

O'Brien, Kevin J., and Lianjiang Li. 1999. "Selective Policy Implementation in Rural China." *Comparative Politics* 31 (2): 167–86.

O'Brien, Kevin J., and Lianjiang Li. 2006. *Rightful Resistance in Rural China.* 1st ed. Cambridge: Cambridge University Press.

O'Hare, Michael. 1977. " 'Not on My Block You Don't': Facility Siting and the Strategic Importance of Compensation." *Public Policy* 25 (4): 407–58.

O'Leary, Rosemary, and Lisa Bingham, eds. 2003. *Promise and Performance of Environmental Conflict Resolution.* Washington, DC: RFF Press.

O'Leary, Rosemary, and Maja Husar. 2002. "What Environmental and Natural Resource Attorneys Really Think About ADR: A National Survey." *Natural Resources & Environment* 16 (4): 262–64.

O'Leary, Rosemary, Tina Nabatchi, and Lisa Bingham. 2005. "Assessing and Improving Conflict Resolution in Multiparty Environmental Negotiations." *International Journal of Organization Theory and Behavior* 8 (2): 181–209.

O'Leary, Rosemary, and Susan Summers Raines. 2001. "Lessons Learned from Two Decades of Alternative Dispute Resolution Programs and Processes at the U.S. Environmental Protection Agency." *Public Administration Review* 61 (6): 682–92.

Orr, Patricia J., Kirk Emerson, and Dale L. Keyes. 2008. "Environmental Conflict Resolution Practice and Performance: An Evaluation Framework." *Conflict Resolution Quarterly* 25 (3): 283–301.

Ostrom, Elinor. 1990. *Governing the Commons: The Evolution of Institutions for Collective Action*. Cambridge: Cambridge University Press.

Ostrom, Elinor. 2005. *Understanding Institutional Diversity*. Princeton, NJ: Princeton University Press.

Ostrom, Elinor. 2007. "Institutional Rational Choice: An Assessment of the Institutional Analysis and Development Framework." In *Theories of the Policy Process*, edited by Paul Sabatier, 21–64. Boulder, CO: Westview Press.

Ostrom, Elinor. 2011. "Background on the Institutional Analysis and Development Framework." *Policy Studies Journal* 39 (1): 7–27.

Ostrom, Elinor, James Walker, and Roy Gardner. 1992. "Covenants with and without a Sword: Self-Governance Is Possible." *The American Political Science Review* 86 (2): 404–17.

O'Toole, Laurence J. 1988. "Strategies for Intergovernmental Management: Implementing Programs in Interorganizational Networks." *International Journal of Public Administration* 11 (4): 417–41.

O'Toole, Laurence J. 1997. "Treating Networks Seriously: Practical and Research-Based Agendas in Public Administration." *Public Administration Review* 57 (1): 45–52.

Parsons, Wayne. 1995. *Public Policy: An Introduction to the Theory and Practice of Policy Analysis*. 1st ed. Cheltenham: Edward Elgar.

Perry, Elizabeth, and Mark Selden, eds. 2010. *Chinese Society: Change, Conflict and Resistance*. 3rd ed. London: Routledge.

Pierre, Jon, ed. 2000. *Debating Governance: Authority, Steering, and Democracy*. 1st ed. Oxford: Oxford University Press.

Pierre, Jon, and Guy Peters. 2000. *Governance, Politics and the State*. New York: Palgrave Macmillan.

Pierson, Paul. 2004. *Politics in Time: History, Institutions, and Social Analysis*. 1st ed. Princeton, NJ: Princeton University Press.

Plummer, Janelle, and John Taylor, eds. 2004. *Community Participation in China: Issues and Processes for Capacity Building*. London: Routledge.

Provan, Keith G., Amy Fish, and Joerg Sydow. 2007. "Interorganizational Networks at the Network Level: A Review of the Empirical Literature on Whole Networks." *Journal of Management* 33 (3): 479–516.

Provan, Keith G., and H. Brinton Milward. 1995. "A Preliminary Theory of Interorganizational Network Effectiveness: A Comparative Study of Four Community Mental Health Systems." *Administrative Science Quarterly* 40 (1): 1–33.

Przeworski, Adam, and Henry J. Teune. 1970. *The Logic of Comparative Social Inquiry*. 1st ed. New York: John Wiley.

Radford, K. J. 1977. *Complex Decision Problems: An Integrated Strategy for Resolution*. 1st ed. Reston, VA: Reston Publishing.

Ragin, Charles. 1987. *The Comparative Method: Moving beyond Qualitative and Quantitative Strategies*. Berkeley, CA: University of California Press.

Ragin, Charles. 1992. "Casing and the Process of Social Inquiry." In *What Is a Case? Exploring the Foundations of Social Inquiry*, edited by Charles Ragin and Howard Becker, 217–26. Cambridge: Cambridge University Press.

Ragin, Charles. 2000. *Fuzzy-Set Social Science*. 1st ed. Chicago: University of Chicago Press.

Ragin, Charles. 2005. "Core versus Tangential Assumptions in Comparative Research." *Studies in Comparative International Development* 40 (1): 33–8.

Ragin, Charles. 2006. "Set Relations in Social Research: Evaluating Their Consistency and Coverage." *Political Analysis* 14 (3): 291–310.

Ragin, Charles. 2008. *Redesigning Social Inquiry: Fuzzy Sets and Beyond.* Chicago: University of Chicago Press.

Register, Richard. 2006. *EcoCities: Rebuilding Cities in Balance with Nature.* Gabriola, BC, Canada: New Society Publishers.

Rhodes, Rod. 1990. "Policy Networks: A British Perspective." *Journal of Theoretical Politics* 2 (3): 293–317.

Rhodes, Rod. 1996. 'The New Governance: Governing without Government'. *Political Studies* 44 (4): 652–67.

Rhodes, Rod. 2007. "Understanding Governance: Ten Years On." *Organization Studies* 28 (8): 1243–64.

Rihoux, Benoît. 2006. "Qualitative Comparative Analysis (QCA) and Related Systematic Comparative Methods: Recent Advances and Remaining Challenges for Social Science Research." *International Sociology* 21 (5): 679–706.

Rihoux, Benoît, and Charles Ragin. 2009. *Configurational Comparative Methods: Qualitative Comparative Analysis (QCA) and Related Techniques.* Thousand Oaks, CA: Sage.

Rittel, Horst W. J., and Melvin M. Webber. 1973. "Dilemmas in a General Theory of Planning." *Policy Sciences* 4 (2): 155–69.

Roberts, Nancy. 1997. "Public Deliberation: An Alternative Approach to Crafting Policy and Setting Direction." *Public Administration Review* 57 (2): 124–32.

Roberts, Nancy. 2004. "Public Deliberation in an Age of Direct Citizen Participation." *The American Review of Public Administration* 34 (4): 315–53.

Roberts, Nancy, and Raymond Trevor Bradley. 1991. "Stakeholder Collaboration and Innovation: A Study of Public Policy Initiation at the State Level." *The Journal of Applied Behavioral Science* 27 (2): 209–27.

Robinson, Scott E. 2006. "A Decade of Treating Networks Seriously." *Policy Studies Journal* 34 (4): 589–98.

Rootes, Christopher, and Liam Leonard. 2009. "Environmental Movements and Campaigns against Waste Infrastructure in the United States." *Environmental Politics* 18 (6): 835–50.

Rothstein, Bo. 2015. "The Chinese Paradox of High Growth and Low Quality of Government: The Cadre Organization Meets Max Weber." *Governance* 28 (4): 533–48.

Sabatier, Paul. 1986. "Top-Down and Bottom-Up Approaches to Implementation Research: A Critical Analysis and Suggested Synthesis." *Journal of Public Policy* 6 (1): 21–48.

Sabatier, Paul. 1988. "An Advocacy Coalition Framework of Policy Change and the Role of Policy-Oriented Learning Therein." *Policy Sciences* 21 (2–3): 129–68.

Sabatier, Paul, eds. 2007. *Theories of the Policy Process.* 2nd ed. Boulder, CO: Westview Press.

Sabatier, Paul, and Hank Jenkins-Smith. 1993. *Policy Change and Learning: An Advocacy Coalition Approach.* Boulder, CO: Westview Press.

Savolainen, Jukka. 1994. "The Rationality of Drawing Big Conclusions Based on Small Samples: In Defense of Mill's Methods." *Social Forces* 72 (4): 1217–24.

Scharpf, Fritz. 1978. "Interorganizational Policy Studies: Issues, Concepts and Perspectives." In *Interorganizational Policymaking: Limits to Coordination and Central Control*, edited by Kenneth Hanf and Fritz Scharpf, 345–70. London: Sage.

Scharpf, Fritz. 1997. *Games Real Actors Play: Actor-Centered Institutionalism in Policy Research*. Boulder, CO: Westview Press.

Schively, Carissa. 2007. "Understanding the NIMBY and LULU Phenomena: Reassessing Our Knowledge Base and Informing Future Research." *Journal of Planning Literature* 21 (3): 255–66.

Schlager, Edella, and Christopher Weible. 2013. "New Theories of the Policy Process." *Policy Studies Journal* 41 (3): 389–96.

Schneider, Carsten, and Ingo Rohlfing. 2013. "Combining QCA and Process Tracing in Set-Theoretic Multi-Method Research." *Sociological Methods & Research* 42 (4): 559–97.

Schneider, Carsten, and Claudius Wagemann. 2010. "Standards of Good Practice in Qualitative Comparative Analysis (QCA) and Fuzzy-Sets." *Comparative Sociology* 9 (3): 397–418.

Schneider, Carsten, and Claudius Wagemann. 2012. *Set-Theoretic Methods for the Social Sciences: A Guide to Qualitative Comparative Analysis*. Cambridge: Cambridge University Press.

Schneider, Mark, John Scholz, Mark Lubell, Denisa Mindruta, and Matthew Edwardsen. 2003. "Building Consensual Institutions: Networks and the National Estuary Program." *American Journal of Political Science* 47 (1): 143–58.

Schön, Donald A., and Martin Rein. 1994. *Frame Reflection: Toward the Resolution of Intractable Policy Controversies*. New York: Basic Books.

Schwartz, Jonathan. 2004. "Environmental NGOs in China: Roles and Limits." *Pacific Affairs* 77 (1): 28–49.

Seawright, Jason, and John Gerring. 2008. "Case Selection Techniques in Case Study Research: A Menu of Qualitative and Quantitative Options." *Political Research Quarterly* 61 (2): 294–308.

Shi, Fayong, and Yongshun Cai. 2006. "Disaggregating the State: Networks and Collective Resistance in Shanghai." *The China Quarterly* 186: 314–32.

Skocpol, Theda. 1979. *States and Social Revolutions: A Comparative Analysis of France, Russia and China*. Cambridge: Cambridge University Press.

Snow, David A., and Robert D. Benford. 1992. "Master Frames and Cycles of Protest." In *Frontiers in Social Movement Theory*, edited by Morris Aldon and Mueller Carol, 133–55. New Haven, CT: Yale University Press.

Sørensen, Eva. 2006. "Metagovernance: The Changing Role of Politicians in Processes of Democratic Governance." *The American Review of Public Administration* 36 (1): 98–114.

Sørensen, Eva, and Jacob Torfing. 2003. "Network Politics, Political Capital, and Democracy." *International Journal of Public Administration* 26 (6): 609–34.

Sørensen, Eva, and Jacob Torfing. 2005. "The Democratic Anchorage of Governance Networks." *Scandinavian Political Studies* 28 (3): 195–218.

Sørensen, Eva, and Jacob Torfing, eds. 2007. *Theories of Democratic Network Governance*. Basingstoke: Palgrave Macmillan.

Sørensen, Eva, and Jacob Torfing. 2009. "Making Governance Networks Effective and Democratic Through Metagovernance." *Public Administration* 87 (2): 234–58.

Stalley, Phillip, and Dongning Yang. 2006. "An Emerging Environmental Movement in China?" *The China Quarterly* 186 (June): 333–56.

Stone, Deborah A. 1989. 'Causal Stories and the Formation of Policy Agendas'. *Political Science Quarterly* 104 (2): 281–300.

Strauss, Anselm, and Juliet M. Corbin. 1990. *Basics of Qualitative Research: Grounded Theory Procedures and Techniques*. 2nd ed. Newbury Park, CA: Sage.

Sullivan, Jonathan, and Lei Xie. 2009. "Environmental Activism, Social Networks and the Internet." *The China Quarterly* 198 (June): 422–32.

Susskind, Lawrence, and Jeffrey Cruikshank. 1987. *Breaking The Impasse.* New York: Basic Books.

Susskind, Lawrence, and Sarah McKearnan. 1999. "The Evolution of Public Policy Dispute Resolution." *Journal of Architectural and Planning Research* 16 (2): 96–115.

Susskind, Lawrence, and Connie Ozawa. 1984. "Mediated Negotiation in the Public Sector: The Planner as Mediator." *Journal of Planning Education and Research* 4 (1): 5–15.

Susskind, Lawrence, Mieke van der Wansem, and Armand Ciccarelli. 2000. *Mediating Land Use Disputes: Pros and Cons.* Cambridge, MA: Lincoln Institute of Land Policy.

Susskind, Lawrence, and Alan Weinstein. 1980. "Towards a Theory of Environmental Dispute Resolution." *Boston College Environmental Affairs Law Review* 9: 311–58.

Tang, Shui-Yan, and Xueyong Zhan. 2008. "Civic Environmental NGOs, Civil Society, and Democratisation in China." *The Journal of Development Studies* 44 (3): 425–48.

Tarrow, Sidney. 1994. *Power in Movement: Social Movements, Collective Action and Politics.* 1st ed. Cambridge: Cambridge University Press.

Teisman, Geert R. 2000. "Models for Research into Decision-Making Processes: On Phases, Streams and Decision-Making Rounds." *Public Administration* 78 (4): 937–56.

Termeer, Catrien J. A. M. 2009. "Barriers to New Modes of Horizontal Governance." *Public Management Review* 11 (3): 299–316.

Thibaut, Johann L. 2011. "An Environmental Civil Society in China? Bridging Theoretical Gaps through a Case Study of Environmental Protest." *Internationales Asien Forum. International Quarterly for Asian Studies* 42 (1–2): 135–63.

Thiem, Alrik, and Adrian Dusa. 2013. "QCA: A Package for Qualitative Comparative Analysis." *The R Journal* 5 (1): 87–97.

Thomas, Gerald B. 1999. "External Shocks, Conflict and Learning as Interactive Sources of Change in U.S. Security Policy." *Journal of Public Policy* 19 (2): 209–31.

Tong, Yanqi. 2005. "Environmental Movements in Transitional Societies: A Comparative Study of Taiwan and China." *Comparative Politics* 37 (2): 167–88.

Tong, Yanqi. 2009. "Dispute Resolution Strategies in a Hybrid System." *China Review* 9 (1): 17–43.

Torfing, Jacob, Guy Peters, Jon Pierre, and Eva Sørensen. 2012. *Interactive Governance: Advancing the Paradigm.* 1st ed. Oxford: Oxford University Press.

Tsang, Steve. 2009. "Consultative Leninism: China's New Political Framework." *Journal of Contemporary China* 18 (62): 865–80.

Tummers, Lars, and Niels Karsten. 2012. "Reflecting on the Role of Literature in Qualitative Public Administration Research: Learning from Grounded Theory." *Administration & Society* 44 (1): 64–86.

van Bueren, Ellen, Erik-Hans Klijn, and Joop F. M. Koppenjan. 2003. "Dealing with Wicked Problems in Networks: Analyzing an Environmental Debate from a Network Perspective." *Journal of Public Administration Research and Theory* 13 (2): 193–212.

van Buuren, Arwin, and Erik-Hans Klijn. 2006. "Trajectories of Institutional Design in Policy Networks: European Interventions in the Dutch Fishery Network as an Example." *International Review of Administrative Sciences* 72 (3): 395–415.

Van Kersbergen, Kees, and Frans van Waarden. 2004. "'Governance' as a Bridge between Disciplines: Cross-Disciplinary Inspiration Regarding Shifts in Governance and Problems of Governability, Accountability and Legitimacy." *European Journal of Political Research* 43 (2): 143–71.

van Rooij, Benjamin. 2010. "The People vs. Pollution: Understanding Citizen Action against Pollution in China." *Journal of Contemporary China* 19 (63): 55–77.

van Rooij, Benjamin. 2012. "The People's Regulation: Citizens and Implementation of Law in China." *Columbia Journal of Asian Law* 25 (2): 116–79.

van Rooij, Benjamin, Rachel E. Stern, and Kathinka Fürst. 2016. "The Authoritarian Logic of Regulatory Pluralism: Understanding China's New Environmental Actors." *Regulation & Governance* 10 (1): 3–13.

van Waarden, Frans. 1992. "Dimensions and Types of Policy Networks." *European Journal of Political Research* 21 (1–2): 29–52.

Vink, Maarten P., and Olaf van Vliet. 2013. "Potentials and Pitfalls of Multi-Value QCA Response to Thiem." *Field Methods* 25 (2): 208–13.

Vis, Barbara. 2012. "The Comparative Advantages of fsQCA and Regression Analysis for Moderately Large-N Analyses." *Sociological Methods & Research* 41 (1): 168–98.

Wagemann, Claudius, and Carsten Schneider. 2010. "Qualitative Comparative Analysis (QCA) and Fuzzy-Sets: Agenda for a Research Approach and a Data Analysis Technique." *Comparative Sociology* 9 (3): 376–96.

Walgrave, Stefaan, Stuart Soroka, and Michiel Nuytemans. 2007. "The Mass Media's Political Agenda-Setting Power: A Longitudinal Analysis of Media, Parliament, and Government in Belgium (1993 to 2000)." *Comparative Political Studies* 41 (6): 814–36.

Walgrave, Stefaan, and Frédéric Varone. 2008. "Punctuated Equilibrium and Agenda-Setting: Bringing Parties Back in: Policy Change after the Dutroux Crisis in Belgium." *Governance* 21 (3): 365–95.

Walgrave, Stefaan, Frédéric Varone, and Patrick Dumont. 2006. "Policy with or Without Parties? A Comparative Analysis of Policy Priorities and Policy Change in Belgium, 1991 to 2000." *Journal of European Public Policy* 13 (7): 1021–38.

Walker, Henry A., and Bernard P. Cohen. 1985. "Scope Statements: Imperatives for Evaluating Theory." *American Sociological Review* 50 (3): 288–301.

Wang, Qing-Jie. 2005. "Transparency in the Grey Box of China's Environmental Governance: A Case Study of Print Media Coverage of an Environmental Controversy from the Pearl River Delta Region." *The Journal of Environment & Development* 14 (2): 278–312.

Warwick, Mara, and Leonard Ortolano. 2007. "Benefits and Costs of Shanghai's Environmental Citizen Complaints System." *China Information* 21 (2): 237–68.

Weible, Christopher, Tanya Heikkila, Peter Deleon, and Paul Sabatier. 2012. "Understanding and Influencing the Policy Process." *Policy Sciences* 45 (1): 1–21.

Weible, Christopher, Paul Sabatier, and Mark Lubell. 2004. "A Comparison of a Collaborative and Top-Down Approach to the Use of Science in Policy: Establishing Marine Protected Areas in California." *Policy Studies Journal* 32 (2): 187–207.

Weiss, Janet A. 1989. "The Powers of Problem Definition: The Case of Government Paperwork." *Policy Sciences* 22 (2): 97–121.

Weller, Robert P. 2012. "Responsive Authoritarianism and Blind-Eye Governance in China." In *Socialism Vanquished, Socialism Challenged*, edited by Nina Bandelj and Dorothy J. Solinger, 83–100. Oxford: Oxford University Press.

Wildavsky, Aaron. 1979. *Speaking Truth to Power: The Art and Craft of Policy Analysis*. 1st ed. Boston, MA: Little, Brown and Company.

Wolsink, Maarten. 1994. "Entanglement of Interests and Motives: Assumptions behind the NIMBY-Theory on Facility Siting." *Urban Studies* 31 (6): 851–66.

Wolsink, Maarten. 2000. "Wind Power and the NIMBY-myth, Institutional Capacity and the Limited Significance of Public Support." *Renewable Energy* 21 (1): 49–64.

Wolsink, Maarten. 2006. "Invalid Theory Impedes Our Understanding: A Critique on the Persistence of the Language of NIMBY." *Transactions of the Institute of British Geographers* 31 (1): 85–91.

Wolsink, Maarten, and Jeroen Devilee. 2009. "The Motives for Accepting or Rejecting Waste Infrastructure Facilities. Shifting the Focus from the Planners' Perspective to Fairness and Community Commitment." *Journal of Environmental Planning and Management* 52 (2): 217–36.

Wondolleck, Julia, and Steven Yaffee. 2000. *Making Collaboration Work: Lessons from Innovation in Natural Resource Management*. Washington, DC: Island Press.

Xie, Lei. 2011. *Environmental Activism in China*. London: Routledge.

Xie, Lei, and Hein-Anton van der Heijden. 2010. "Environmental Movements and Political Opportunities: The Case of China." *Social Movement Studies* 9 (1): 51–68.

Yang, Guobin. 2005. "Environmental NGOs and Institutional Dynamics in China." *The China Quarterly* 181 (March): 46–66.

Yang, Guobin. 2010. "Brokering Environment and Health in China: Issue Entrepreneurs of the Public Sphere." *Journal of Contemporary China* 19 (63): 101–18.

Yin, Robert K. 2008. *Case Study Research: Design and Methods*. 4th ed. Los Angeles, CA: Sage.

Yu, Jianrong. 2007. "Social Conflict in Rural China." *China Security* 3 (2): 2–17.

Zhao, Yuhong. 2004. "Environmental Dispute Resolution in China." *Journal of Environmental Law* 16 (2): 157–92.

Zheng, Haitao, Martin De Jong, and Joop Koppenjan. 2010. "Applying Policy Network Theory to Policy-Making in China: The Case of Urban Health Insurance Reform." *Public Administration* 88 (2): 398–417.

Index